The Blue Andalusian Chicken and the Inheritance of Blue in Poultry

by Professor William A. Lippincott

with an introduction by Jackson Chambers

This work contains material that was originally published in 1921.

This publication is within the Public Domain.

*This edition is reprinted for educational purposes
and in accordance with all applicable Federal Laws.*

Introduction Copyright 2018 by Jackson Chambers

The World's Largest Selection of Vintage Poultry Books

www.VintagePoultry.com

Self Reliance Books

Get more historic titles on animal and stock breeding, gardening and old fashioned skills by visiting us at:

http://selfreliancebooks.blogspot.com/

Introduction

I am pleased to present yet another title on Poultry.

The work is in the Public Domain and is re-printed here in accordance with Federal Laws.

As with all reprinted books of this age that are intended to perfectly reproduce the original edition, considerable pains and effort had to be undertaken to correct fading and sometimes outright damage to existing proofs of this title. At times, this task is quite monumental, requiring an almost total "rebuilding" of some pages from digital proofs of multiple copies. Despite this, imperfections still sometimes exist in the final proof and may detract from the visual appearance of the text.

I hope you enjoy reading this book as much as I enjoyed making it available to readers again.

Jackson Chambers

THE CASE OF THE BLUE ANDALUSIAN[1]

WILLIAM A. LIPPINCOTT

KANSAS AGRICULTURAL EXPERIMENT STATION, MANHATTAN, KANSAS

The blue Andalusian has become the classic in animals as an example of a heterozygote phenotypically intermediate between the parental types. It has also served as an illustration of the failure of dominance for those opponents of Mendelism who consider dominance one of its fundamentals. Furthermore, it has been in constant demand as a classroom example of blended inheritance.

The main facts concerning the breeding behavior of blue Andalusians are, accordingly, more or less familiar. In spite of the long-continued efforts of their breeders they do not come true to color as a breed, but continually throw a certain proportion of off-colored progeny, or "wasters," of two kinds. One is self (entirely) black. The other approaches white, but displays considerable pigment, and is referred to variously as white, splashed, and splashed-white. Since an examination of a large number of birds of this type shows the pigmented feathers to be *blue* in all sections of the female and in those sections of the male which carry blue feathers in the blue Andalusian male, they will be referred to throughout this paper as blue-splashed.

"Splashed" refers to the fact that the pigment does not regularly appear in any particular group of feathers or in any definite region. Feathers located apparently at random on any part of the body may be pigmented over their entire surface or may show only slight traces of pigment. Not infrequently both of these conditions are present in the same individual.

Since the blacks and blue-splashed breed true when

[1] Contribution from the Department of Experimental Breeding, Wisconsin Agricultural Experiment Station, No. 12, and from the Kansas Agricultural Experiment Station.

mated *inter se,* they are considered as being homozygous. If crossed they invariably produce blues.

These facts have led to the current view that the case involves a single allelomorphic pair of characters. The blacks and blue-splashed represent the homozygous conditions, while the self blue is the heterozygote between the two. When blues are interbred, blacks, blues, and blue-splashed are produced in a ratio approximating 1:2:1 for these classes, respectively, which seems to corroborate this view.

Although the blacks and blue-splashed breed true for color, they are not recognized by fanciers as breeds or varieties, and it is doubtful whether they would continue to exist if, much to the disgust of the breeders of blue Andalusians, they did not continue to appear as "wasters" among the progeny of blues. The blues, on the other hand, are quite widely bred. They are officially recognized by the American Poultry Association as a distinct breed and have their place in the American Standard of Perfection. It is interesting in this connection to note that the numbers of blues they throw on the Mendelian expectation barely gets them into the Standard, since the rules of the Association are that no breed can be officially recognized as such unless a minimum of 50 per cent. of the offspring come reasonably true to type (American Poultry Association, 1910, p. 328, Constitution, Article XI).

The blues are quite uniformly bluish-gray throughout the body, with certain exceptions in the males to be noted later. Emphasis has usually been laid on their distinctness from the black and the blue-splashed birds, but it seems important to note their resemblance to these two classes. In the first place, they are like the blacks in being *self*-colored, that is, all feathers in all parts of the body are pigmented. In the second they resemble the blue-splashed in that the color of the individual pigmented feathers is *blue* rather than black, save in certain sections of the males of both classes, where the feathers showing pigment are a glossy black, apparently a secondary sexual characteristic. The blue appearance is due to the distribution and arrangement of the pigment granules in the feather structure, as will be described later. The fact that the splashed birds are splashed with blue (with the exception noted above) rather than black is important and appears not to have been noted, or at least not emphasized, by previous writers.

As an example, Punnett (1911, p. 70), in discussing the breeding behavior of the blue Andalusian, says: "It always throws 'wasters' of two kinds, viz., blacks, and whites *splashed with black*" (italics mine). In the material which has come under my observation, consisting of upwards of one hundred birds in the unrelated flocks of the poultry departments of Kansas State Agricultural College and the University of Wisconsin, no individual has been noted in which the pigmented areas were not distinctly bluish-gray, except that those pigmented feathers or parts of feathers appearing in the hackle, back, and saddle of the male were glossy black. These sections, it should be clearly understood, are also glossy black in the blue Andalusian male. There are occasionally flecks or small spots of black, appearing in the blue-gray feathers, and even in the white feathers of the blue-splashed birds. This is also true of the blues and, indeed, is not a rare occurrence in both dominant and recessive white races of other breeds. It does not in the least affect the fact that, in the material so far observed, the white birds have been splashed with bluish-gray rather than black in those sections where the blue Andalusian is also blue. This conclusion is borne out by the results of a microscopic examination.

In an effort to determine the fundamental differences between the three Andalusian phenotypes, a careful study of feathers from numerous individuals of each phenotype was made. A detailed account of the results of the study will be published in a later paper. For present purposes a short account of the most obvious differences will serve.

The pigment in all three phenotypes is black. The differences in appearance are due to the distribution and arrangement of the pigment or to its absence.

The pigment in a black Andalusian feather is in the form of rod-shaped granules, which almost completely fill each cell. They extend to the very tips of both curved and hooked barbules, and into the tiny hooklets given off from the barbs of the latter class. The cell boundaries are usually visible, due, apparently, to a slight contraction of the pigment, leaving very narrow pigment-free spaces between the cells. The former position of the nucleus of each cell is almost always plainly visible, due to an accumulation of pigment at its border, and to a narrow area surrounding it that bears relatively little pigment. In appearance, size and distribution the pigment granules in feathers from the black Langshan seem to be identical with those of black Andalusians.

The feathers from blue Andalusians differ from those of the blacks in two important particulars, namely, the restriction of the pigment in the feather structure and the shape of the granules. In blues of average shade, pigment fails to appear in the extremities of the barbules of both types. The hooklets are also entirely pigment-free. Though not always the case, the curved barbules usually carry rather more pigment than the hooked barbules, since the pigment extends further toward the distal end. As a usual thing that part of the hooked barbule which bears the hooks is free from pigment and does not differ in appearance, by transmitted light, from the same portion of a similar barbule from a white feather.

In the pigmented portions the pigment is usually markedly contracted or clumped within each cell, leaving a pigmentless space about the border much wider than is the case with blacks. These spaces are not always clean cut, but may be broken by invading rows of granules, or isolated granules may be found scattered within them. As a usual thing the nuclear boundaries in the cells of blue-gray feathers can only be made out with difficulty, if at all.

In cross-section, the pigment granules are seen to be scattered through the cortex of the barb and along the boundaries of the medullary cells. They are not restricted to the apex of the barb, as is reported by Lloyd-Jones (1915, p. 472, Figs. 37–39) in the so-called blue pigeon.

The predominating shape of the pigment granules in feathers from blue Andalusians is round. There may be a few elliptical granules and occasionally one which can not be classified otherwise than as a rod. These are quite rare, however, and one may carefully scrutinize several blue-gray feathers without finding any but round or very slightly elliptical granules. These round granules quite frequently appear in straight rows, giving the effect of a string of beads.

While the granule shape may have an appreciable effect in giving the bluish-gray cast found in blues and blue-splashed, it seems more likely that, as suggested above, the bluish appearance is due to the restriction or arrangement of the pigment. While the condition is not precisely the same as in pigeons, as described by Cole (1914, pp. 324–325) and Lloyd-Jones (1915, pp. 472–473), the optical effect appears to be from essentially the same causes, namely, the clumping of the pigment within the cells, and the reflection from this pigment through more or less transparent layers of keratin. It appears, however, that in the blue Andalusian the contrast between the pigment-free ends of the barbules and the pigmented barbs and barbule-bases is of more importance in producing the bluish effect than is suggested for the pigeon by these writers.

A characteristic of the typical blue Andalusian not before mentioned is that the contour feathers on the female and the breast feathers on the male present a laced appearance. This results from a black edging on that portion of each feather which is exposed when in its natural position. In this part of the feather the barbules on both sides of the barb are alike, being without hooks. The cells in these barbules are more heavily pigmented than is true of the rest of the feather and the granules are rod shaped. In the regions where the black is giving way to blue, both round and rod-shaped granules are found.

All pigmented feathers secured from several blue-splashed females show identically the same pigment arrangement and granule shape as predominates in the blues. This holds true whether the portion examined comes from a feather that is pigmented throughout, or from one that is almost wholly white, with but a trace of pigment showing. In feathers which are pigmented throughout, the same relation regarding the lacing occurs as in homologous feathers in blue females.

The statements of the foregoing paragraph apply equally well to the feathers of those sections of the blue-splashed male which are blue in the blue male.

As previously mentioned, in both blue-splashed and blues, as well as in other self-colored races, black flecking or spotting not infrequently appears. Such spots, whether taken from a blue feather from a blue individual, or from a blue or an almost white feather from a blue-splashed bird, invariably show rod-shaped granules, while the surrounding area, if blue-gray, shows round granules. These spots are apparently entirely independent of the factors and conditions discussed in this paper and their appearance is comparatively limited. If hereditary, they probably depend on other factors. In handling blues and blue-splashed, however, one can not help being impressed with the possibility that these spots are caused by some interference with the full expression of the factors responsible for the arrangement and rounding of the pigment granules. Whether this interference is hereditary or environmental is as yet undetermined.

One further fact concerning the blue Andalusian males, already alluded to, is of interest. The long feathers of the neck (hackle) and saddle are glossy black. This is apparently a secondary sexual characteristic, though it is as yet undetermined whether it is due to the presence of testicular secretion or the absence of ovarian secretion. The black feathers from both sections show rod-shaped granules predominating. There are numerous elliptical granules and a few round granules present. The pigment is not restricted as to distribution in the feather structure and is found even in the tiny hooklets of the hooked barbules, being in all these respects similar to the analogous feathers on a black male. These same conditions prevail in homologous pigmented feathers in a blue-splashed male.

The foregoing describes the conditions that usually prevail. There is some variation in all conditions described. In pure-bred blue Andalusians, for instance, there frequently appear areas that are not the usual clear blue-gray, but are dull and smoky. In such regions both round and rod-shaped granules are found in about equal numbers.

Bateson and Punnett (1906, p. 20) make note of the fact that the adult color of Andalusians may be determined from the down color of the young chicks. Examinations of the down show the same differences in granule shape that are observed in the adults. The blue and blue-splashed chicks for the most part show nothing but round granules in the down, while the blacks show rods.

It is of interest to note in this connection that a section from that portion of a barred Plymouth Rock feather where the black bar is giving way to the white, and the color is dull gray or dun with no bluish cast, there is a dilution of pigment as to amount, but no restriction as to arrangement or distribution. The pigment is fully extended through the barbule cells and consists of rod-shaped granules. There simply appears to be less pigment. While this is the usual condition, here, too, there is variation. At least one barred Rock individual was found whose feathers showed numerous round granules, though the rods predominated.

While it is generally accepted that blue Andalusians, when mated *inter se,* produce blacks, blues and blue-splashed in the ratio of 1 black to 2 blues to 1 blue-splashed, exact data on this mating, as well as on the back crosses to black and blue-splashed, are really very meager. Bateson and Saunders (1902, p. 131) first suggested that the blue Andalusian was probably a heterozygote. Bateson and Punnett (1905, p. 118) quoted Mrs. Blacket Gill, a fancier of blue Andalusians, to the effect that blues mated to blues gave 22 blacks, 36 blues and 17 white-splashed (*i. e.*, blue-splashed). They secured stock from Mrs. Gill and made matings which gave the following results:

By the blue ♂ the white ♀ gave 34 blue, 20 white-splashed, and the black ♀ gave 27 blue, 19 black. In each case the result is qualitatively what would be expected if the blue is a heterozygote of *black × splashed white* [italics mine]; but whether the departure from equality indicates that some gametes bear the unsegregated blue, or may merely be taken as individual irregularities, can not yet be stated.

The same blue cock was bred with a black hen from Experiment 40 (in which the dark birds were unexpected), F_2, from White Wyandotte × Wh. Legh., giving as offspring 10 black, 15 slaty black to bluish. Hence, therefore, it is evident that the black ♀ was a homozygous black. The 10 blacks are the result of the union of the black gametes from the Andalusian ♂ with those of the ♀, and the 15 slaty resulted from the meeting of the black of the hen with the white-splashed from the Andalusian.

Bateson and Punnett (1906, p. 20) give the following summary of the data upon which the case of the blue Andalusian largely rests at the present time.

In Report I it was suggested that the blue colour of the Andalusian is probably heterozygous, and in Report II (p. 118) figures were given in support of this view. During the past two years additional evidence has been acquired, and every form of mating has now been tested, with the following results:

No. of Experiment	Nature of Mating	Result		
		Black	Blue	Wh. Spl.
Rep. II, p. 118.....	Blue ♀'s × blue ♂.............	22	36	17
Exp. 276...........	Blue ♀'s × blue ♂.............	19	42	22
(Total numbers for blue and blue.....................		41	78	39)
[Expectation (inserted by the writer)....................		39.5	79	39.5]
Rep. II, p. 118.....	Wh. spl. ♀ × blue ♂............	—	34	20
" " " "	Black ♀ × blue ♂..............	19	27	—
Exp. 269..........	Wh. spl. ♀'s × wh. spl. ♂........	—	—	40
" 270..........	Black ♀ × wh. spl. ♂...........	—	20	—
" 294..........	Black ♀ × black ♂.............	25	—	—

The colour of most of the chickens was determined in the down. In the blacks the down is black with the exception of the ventral surface, the tips of the wings, and sometimes parts of the head, which are white. The down in the blues is slaty-blue, similarly marked with white, whilst in the white splashed it is of an exceedingly pale blue tint as a rule, though sometimes practically colourless.

The above figures bear out the view we previously expressed as to the heterozygous nature of the blues, . . .

The only other definite figures that have come under the writer's notice are from W. J. Coates, a blue Andalusian breeder of East Calais, Vermont, quoted by Platt (1916, p. 665) and referred to by Pearl (1917, p. 149). These are for matings of blue to blue and are as follows:

Mating	White (Blue-Splashed)	Blue	Black	Dark Red
A	4	10	3	1
B	4	5	2	0
C	3	3	0	3
D	0	12	1	0
E	3	3	1	0
	14	33	7	4

The fact that birds showing dark red appear is unusual and would seem to indicate that the Coates stock differs in its genetic constitution from the majority of the members of the breed, unless the occasional appearance of red is a fact usually suppressed by breeders.

Bateson and his co-workers make no attempt beyond that quoted above to account for the hereditary behavior of Andalusians and appear content to rest the case on the assumption that "blue is a heterozygote of black × splashed white."

The fact that "blue" is not a true intermediate between black and blue-splashed does not seem to have received due consideration. While the blue-gray bird is in a sense intermediate between self black and an individual that approaches white more or less closely, this intermediacy is more apparent than real. As previously pointed out, it is not intermediate in regard to either of the conditions involved when they are considered separately. It resembles the black phenotype in being self-colored and the blue-splashed phenotype in having the pigment restriction within the barbules, which gives the blue-gray effect.

The 1:2:1 ratio may therefore be analyzed as follows:

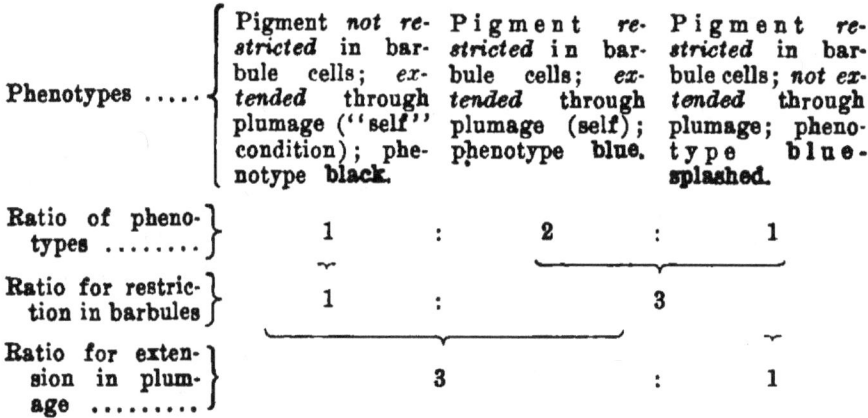

In reality, then, the 1:2:1 ratio is the result of the combination of two 3:1 ratios.

The foregoing facts appear to lend themselves equally well to two interpretations. The first is that there are two pairs of allelomorphic factors at work. The second, that there is one pair of true allelomorphs (*i. e.*, factors having identical loci on homologous chromosomes), neither of which is recessive to the other in its manifestation in the phenotype.

The suggestion of two pairs of allelomorphic factors to explain the case of the Andalusian is not a new one. Goldschmidt (1913, p. 274) makes such a suggestion. After pointing out that the offspring of a pair of blues are black, blue, and "*schmutzigweiss*" in the ratio of 1:2:1, and that all three phenotypes carry pigment, he proposed two factors to account for the condition. The one is an "*Entfaltungsfaktor,*" which brings about a full development of the pigment. He represents this factor by "*Q*" (Quantität) which is possessed by the black race. The other factor, which is possessed by the "Weisse" race, he calls a "*Mosaikfaktor,*" which finely divides the pigment. This factor he designates M (Mosaik). He finds it necessary to postulate further that Q is closely linked with m, and M is closely linked with q. Assuming pigment (P) to be present in all cases, he represents the "black" gamete as (mPQ), the "white" gamete as (MPq), and the F_1 blues as (mPQ)(MPq). The blue results from bringing M and Q into the same zygote. The monohybrid ratio results when the blues are inbred, however, because of the close coupling of the factors within the parentheses.

The Hagedoorns (1914, p. 179) also make use of two coupled factors in accounting for the hereditary behavior of blue Andalusians. They state:

> A blue Andalusian fowl, when mated by us to "recessive" white hens did not produce as many blue as white chicks, as should result on the hypothesis, that the white Andalusian is a recessive white (blue and black Andalusians being heterozygotes and homozygotes for one single genetic factor), but exclusively blacks and blues in equal proportions.

To account for this result they propose a gene A which is present in black Andalusians, but absent in the "white" Andalusian. The blacks, conversely, lack a gene B which is present in the "whites."

> This factor B, present in a pigmented fowl, actively "dilutes" the colour. It has no effect in the white Andalusians, because these, as they lack A, *are not pigmented* [italics mine]. We should therefore expect dilute black (blue) young from the cross black $\times$ white, which, inter se, would give AB, Ab, aB and ab offspring. Now, there is no evidence that in *Andalusians* there are ever produced $aabb$ animals, or $AABB$. There seems to be a mutual repulsion between A and B, so that no AB or ab gametes are ever produced. In some varieties of fowls this repulsion does not seem to exist, as pure strains of blue chickens occur.

Unless their material differs from any that has come under my observation the Hagedoorns err in assuming that what is frequently termed "the white Andalusian" carries no pigment, and Goldschmidt's suggestion accords more closely with the facts. Further, if the "recessive" white to which they refer was an Andalusian, the production of equal numbers of blues and blacks from a blue $\times$ white (blue-splashed) cross is difficult to understand. The expectation would be equal numbers of blue-splashed and blues. If, as I suspect, the "recessive" white was a true recessive from another race, their results can only be interpreted by assuming that the "white" gametes as well as "black" gametes produced by the blue fowl carried a factor necessary for pigment production, which was lacking in the recessive whites.

If the latter is the case it accords with results I have obtained the past season. Among several matings made, preliminary to a further study of Andalusian blue, a white Wyandotte ♂ (*R 840* from the University of Wisconsin flock) was mated with blue-splashed Andalusian ♀♀ *M 409* and *M 539* (also kindly furnished by the poultry department of the University of Wisconsin). From *M 409* seven chicks were hatched, all of which were unmistakably bluish-gray. Six chicks which failed to hatch, but which did develop far enough for the color of the down to be determined, were also all blues. From *M 539*, brought in late in the season with the hope of increasing the numbers of chicks from this type of mating, three chicks were secured, which were again all bluish-gray. On the assumption that Wyandotte white is recessive (I am surprised to find no statement to this effect in the literature) these results would seem to indicate that a factor necessary for pigment formation as well as one causing the characteristic arrangement or restriction of the pigment found in blues, and both lacking in the Wyandotte, were furnished by the blue-splashed Andalusian. And further that a factor for the extension of this pigment to all feathers on the body was furnished by the Wyandotte. The blue offspring from this mating are assuredly not intermediates between a pure white parent and one that appears to be nearly white.

It is significant to note in this connection that the blue-gray offspring of the white Wyandotte × blue-splashed Andalusian cross show pigment granules that are predominatingly round. In some individuals they all appear to be round, while in others some rods may be made out. The down of black chicks, offspring of a blue Andalusian ♂ and a white Plymouth Rock ♀, showed only rod-shaped granules. Feathers from a blue-gray individual, whose dam was a blue-splashed Andalusian and whose sire was a crossbred, the offspring of a Houdan ♂ × single-combed white Leghorn ♀ cross, showed only round granules.

If, as Goldschmidt assumes, his factors mQ and Mq are so closely linked that they never separate, and behave only as a single pair of factors, it is simpler to assume that there is but one pair of factors. As already pointed out, however, the discontinuity in the gradations from blue-splashed to black is such as to lead one strongly to suspect that two pairs of factors are at work. This discontinuity is greatly emphasized in the case of the blue offspring from the white Wyandotte × blue-splashed Andalusian cross. It is perhaps not impossible that a single pair of factors should bring about the result found in Andalusians, but it is so unusual as to make the assumption of two pairs of factors reasonable.

If this assumption is correct it must be further assumed, as Goldschmidt implies but does not state, that the black and splashed races each contribute a dominant and a recessive factor, and that in the blues we have the expression of both dominants, namely, the extension of pigment to all feathers, furnished by the black (or, in the Wyandotte cross noted above, by the white) parent, and the restriction of the pigment in the feather structure in such a way that the effect is bluish-gray, furnished by the blue-splashed parent. It is of interest in this connection to note that the blue condition produced by the restriction of the pigment in the barbule cells is recessive in pigeons (Cole, 1914, p. 325), while in Andalusians, on the above assumption, it is dominant.

While exact data concerning the breeding behavior of blue Andalusians are exceedingly meager, the experience of breeders generally seems to be in accord with such data as there are, and with the interpretation offered by Bateson and his associates. In order to account for the failure to secure a dihybrid ratio from the mating of blues, one is driven to assume linkage, and apparently a quite close linkage, of the dominant of one allelomorphic pair to the recessive of the other. As I hope to make clear, however, the linkage may not be complete, since it would easily be possible for crossing-over to occur occasionally with very slight likelihood of detection.

Goodale (1917, p. 213) has very recently shown that crossing-over occurs in the sex chromosome of the male fowl, though he has not as yet presented his evidence in detail. The universality of the laws of heredity throughout the plant and animal kingdoms is such that it would be a matter of surprise if crossing-over in fowls did not also occur in chromosomes other than the sex chromosome.

There is at present no certain criterion by which to predict whether, having assumed crossing-over in the autosomes of fowls, it occurs in one sex only or in both; and if in but one sex, which it may be, unless one chooses to suppose that it occurs only in the sex homozygous for the sex chromosome, as in Drosophila. If it occurs in both sexes, it is apparently so rare an event in Andalusians that the probability of securing two cross-over individuals in a mating made for purposes of analysis is so small as to be almost negligible.

After a somewhat extended microscopic study of blue-gray feathers from blues, blue-splashed and certain crosses, it seems more in accordance with their apparent action to refer to the factor responsible for changing black into bluish-gray as a restrictor, designated as R rather than M (Mosaikfactor) as was done by Goldschmidt. Similarly in place of Q (Quantität) I would suggest E, as responsible for the extension of pigment to all the feathers of the body.

Using this terminology and assuming for the moment complete linkage, a cross between individuals of the black and blue-splashed races, respectively, would appear as follows:

$$Er\ Er = \text{black} \times eR\ eR = \text{blue-splashed};$$
$$F_1 \qquad Er\ eR = \text{blue};$$

F$_2$ 1 *Er Er* = black + 2 *Er eR* = blue + 1 *eR eR* = blue-splashed.

The gametes produced by the F$_1$ (blues) are *Er* and *eR*. If crossing-over should occur there would be occasional *ER* and *er* gametes produced. It is highly interesting to note that if these two classes of cross-over gametes were produced in equal numbers, as would be expected, and the individuals producing them were mated with ordinary blues, exactly the same phenotypic ratio would result as from the unions of the non-cross-over gametes, viz.:

F$_1$ crossover gametes   Ordinary gametes of F$_1$ blue
 ER, *er*   *Er*, *eR*

F$_2$ *ER Er* = blue, *ER eR* = blue, *er Er* = black, *er eR* = blue-splashed.

This is the usual ratio of 1 black to 2 blues to 1 blue-splashed and would, from the very nature of the case, escape observation as involving crossing-over unless careful analysis were made of the hereditary constitution of these particular F$_2$ individuals.

Such analyses would not be impossible, though they might be long and tedious. The matings which would uncover any of the cross-over types, if offspring were produced in sufficient numbers to make it fairly certain that one were not dealing with chance variations in the ratios, are given herewith.

Cross-over blue of *ER Er* constitution mated with an ordinary blue would give the following expectation:

 ER Er = blue cross-over × *Er eR* = ordinary blue;

F$_1$ *ER Er* = blue,
       ER eR = blue,
       Er Er = black,
       Er eR = blue,

or 3 blues to 1 black, while the ordinary blues would give the normal 1 black to 2 blues to 1 blue-splashed.

Similarly this same individual mated with ordinary blue-splashed would produce all blues instead of the ordinary expectation of 1 blue to 1 blue-splashed, viz.:

$ER\ Er$ = blue cross-over $\times$ $eR\ eR$ = ordinary splashed;

F$_1$ $ER\ eR$ = blue,
      $Er\ eR$ = blue..

If blue cross-over of the type $ER\ eR$ were mated with ordinary black the expectation would be all blues instead of the usual blues and blacks in equal numbers, viz.:

 $ER\ eR$ = blue cross-over $\times$ $Er\ Er$ = black;

F$_1$ $ER\ Er$ = blue,
      $eR\ Er$ = blue.

This second type of blue cross-over individual, $ER\ eR$, mated with ordinary blue, would give an expectation of 3 blues to 1 blue-splashed instead of the ordinary 1:2:1 ratio, viz.:

 $ER\ eR$ = blue cross-over $\times$ $Er\ eR$ = ordinary blue;

F$_1$ $ER\ Er$ = blue,
      $ER\ eR$ = blue,
      $eR\ Er$ = blue,
      $eR\ eR$ = blue-splashed.

If black cross-over $Er\ er$ were mated with ordinary blue the expectation would be 2 blacks to 1 blue to 1 blue-splashed instead of the ordinary ratio of equal numbers of blues and blacks, viz.:

 $Er\ er$ = black cross-over $\times$ $Er\ eR$ = ordinary blue;

F$_1$ $Er\ Er$ = black,
      $Er\ eR$ = blue,
      $er\ Er$ = black,
      $er\ eR$ = blue-splashed.

This same individual $Er\ er$ (black cross-over) mated with an ordinary splashed bird would give an expectation of half blues and half blue-splashed instead of all blues, as in the case of ordinary black and blue-splashed, viz.:

 $Er\ er$ = black cross-over $\times$ $eR\ eR$ = ordinary blue-splashed;

F$_1$ $Er\ eR$ = blue,
      $er\ eR$ = blue-splashed.

Finally, blue-splashed cross-over *eR er* mated with ordinary blue would give an expectation of 1 black to 1 blue to 2 blue-splashed instead of the ordinary expectation of equal numbers of blues and blue-splashed, viz.:

eR er = blue-splashed cross-over × *Er eR* = ordinary blue;

F_1 *eR Er* = blue,
                 eR eR = blue-splashed,
                 er Er = black,
                 er eR = blue-splashed.

The possible matings not indicated in the foregoing are those which would produce the same phenotypic ratios as if ordinary individuals (*i. e.*, non-cross-overs) of the same appearance as the cross-overs were used. Such matings are naturally of no value for analysis.

If it should later be shown that crossing-over does occur as suggested above and there are two pairs of factors concerned, there is the possibility of occasionally securing *ER* gametes. This in turn would seem to make possible the blue Andalusian breeder's long-time dream of producing blues that "breed true." With the appearance of the double recessive gamete *er* another race of Andalusian would apparently become possible, which, if the factors assumed in this paper are correct, should be white splashed with *black* instead of with blue.

The second possible interpretation of the facts so far established is that my postulated factors *R* and *E* occupy identical loci on homologous chromosomes, neither being recessive to the other in its phenotypic expression. For the present at least any evidence that this is the correct interpretation will be largely negative and come from continued failure to find cross-over individuals with regard to *R* and *E*. If these cross-overs should not be found it might at first appear that the interpretation of the case of the blue Andalusian is in all probability exactly what has been suggested from the first, namely, that blue is a heterozygote intermediate between the parental types.

Such an interpretation makes the characters *black* and *blue-splashed* the allelomorphs.

The practise of referring to *characters* that seem to behave in an alternative relationship in heredity as allelomorphs, instead of *factors occupying identical loci on homologous chromosomes,* is, it is to be hoped, passing. That it has lead to a misinterpretation in the present case is shown by the fact that all the offspring of certain pure white birds mated with blue-splashed ones are blue. The E factor must have come from an individual that was homozygous for it and devoid of pigment. It appears reasonable to expect that among the F_2's from the white Wyandotte $\times$ blue-splashed Andalusian cross will appear pure whites that carry the R factor. If this proves to be the case the allelomorphs are two factors, R and E, which act on black pigment. R arranges and restricts the pigment in the feather structure so that it gives a bluish-gray appearance. E extends any black pigment present to all the feathers of the body. One and probably either or both may be present without any phenotypic expression whatsoever. In fact, for every sixteen F_2 individuals from this cross four pure whites are to be expected in which the genotypic ratio with regard to R and E is $1:2:1$, exactly as in the F_2's from a cross of a black and a blue-splashed Andalusian. One of these whites will be homozygous for R like the blue-splashed Andalusian. One will be homozygous for E like the black Andalusian. And two will be heterozygous for E and R, as are the blue Andalusians. But because there is no black pigment present these differences in the genotype do not affect the phenotype. For the sake of clearness the expectation of this cross is shown herewith, carried through the F_2 generation. P is taken to represent a factor necessary for the formation of pigment which is present in the blue-splashed Andalusian, but absent in the white Wyandotte, while E and R are represented as allelomorphic to each other.

White Wyandotte ♂ × blue-splashed Andalusian ♀.

	$ppEE$	$PPRR$		
F_1		$PpRE$ = all blue;		
F_2	6 blues:	3 blue-splashed:	3 black:	4 white.
	2 $PPRE$	1 $PPRR$	1 $PPEE$	1 $ppRR$
	4 $PpRE$	2 $PpRR$	2 $PpEE$	2 $ppRE$
				1 $ppEE$

This same ratio (6:3:3:4), which is to be expected on either interpretation, has been reported by Baur (1914, p. 95) for crosses between a white-flowered race and certain plants bearing ivory-colored flowers, of the snapdragon (*Antirrhinum majus*).

Recessive mutations are of comparatively frequent occurrence. Dominant mutations, though much less frequent, have been described so often that they can not be reasonably doubted. There appears to be no reason, *a priori*, why a mutation might not occur where the mutated factors' potency of expression in the phenotype is approximately equal to that of the normal factor. That this has occurred, not once, but several times, might be the interpretation placed on the striking allelomorphic series reported by Nabours (1914, p. 141) for the color patterns of the grouse locust (*Paratettix*).

Upon which of the two alternative interpretations is correct appears to depend the possible success or the futility of the search for true breeding blues. The first makes it possible. The second appears to close the door of hope in the Andalusian breeder's face unless hope is seen in the progressive selection of the darker blue-splashed individuals. It does not appear possible, on the basis of present known facts, to reach a conclusion. Extensive matings are being made for the coming breeding season which it is hoped will throw further light on the matter.

Summary

1. This paper shows that blue Andalusians are like black Andalusians in that they are self-colored. They

are like the blue-splashed in that homologous pigmented feathers in both sexes have the same condition with reference to the restriction of pigment in the feather structure.

2. The fundamental phenotypic differences between black, blue and blue-splashed Andalusians are briefly described.

3. It is pointed out that the 1:2:1 ratio is in reality a combination of two 3:1 ratios.

4. The condition in the blues is shown to be due to the combined action of two factors R and E. R acts on black pigment, restricting its distribution in such a way that it gives the characteristic blue-gray appearance. E extends black pigment to every feather on the fowl's body.

5. It is impossible to decide on the basis of present facts whether R and E are located on identical loci of homologous chromosomes or are the dominants of two pairs of factors, each linked to the recessive allelomorph of the other.

6. It is shown that if the latter is the condition, crossing-over might occasionally occur between R and E with small likelihood of detection.[2] If crossing-over does occur, RE gametes are possible, which appears in turn to make possible true-breeding blues.

Acknowledgments

It is a pleasure to acknowledge my indebtedness to Dr. Leon J. Cole and Professor Jas. G. Halpin, of the University of Wisconsin. The blue Andalusian problem was undertaken at Dr. Cole's suggestion and the work is being continued under his direction. I have consulted him freely during the preparation of this paper. Pro-

[2] In ordinary practice poultry breeders make what are known as "pen matings," that is, one male is mated to a number of females and the offspring from these females are not kept separate. The exact parentage of any individual is therefore known only with regard to its sire, since its dam might be any one of the females in the group. As the detection of crossing-over depends upon the results of individual matings, it would be practically impossible to discover it under these conditions.

fessor Halpin very kindly placed stock and equipment at my disposal for an entire breeding season, without which it would not have been possible to carry on the breeding work reported herein.

BIBLIOGRAPHY

American Poultry Association.
 1910. The American Standard of Perfection. 331 pp. Pub. by Amer. Poultry Assoc.

Bateson, W., and Saunders, E. R.
 1902. Reports to the Evolution Committee of the Royal Society, I, 160 pp.

Bateson, W., and Punnett, R. C.
 1905. Reports to the Evolution Committee of the Royal Society, II, pp. 99–131.
 1906. Reports to the Evolution Committee of the Royal Society, III, pp. 11–23.

Baur, E.
 1914. Einführung in die experimentelle Vererbungslehre. 2. neubearbeitete Auflage. viii + 401 pp. Berlin: Gebrüder Borntraeger.

Cole, L. J.
 1914. Studies on Inheritance in Pigeons: I. Hereditary Relations of the Principal Colors. R. I. Agr. Expt. Sta., Bull. 158, pp. 311–380, pls. 1–4.

Goldschmidt, R.
 1913. Einführung in die Vererbungswissenschaft. Zweite Auflage, xii + 546 pp. Leipzig: W. Engelmann.

Goodale, H. D.
 1917. Crossing-over in the Sex Chromosome of the Male Fowl. *Science*, N. S., Vol. 46, No. 1183, p. 213.

Hagedoorn, A. L., and A. C.
 1914 Studies on Variation and Selection. *Zeitsch. f. Ind. Abstam. u. Vererbungslehre*, Vol. 11, No. 3, pp. 145–183.

Lloyd-Jones, O.
 1915. Studies on Inheritance in Pigeons: II. A Microscopical and Chemical Study of the Feather Pigments. *Jour. Exp. Zool.*, Vol. 18, No. 3, pp. 453–495, pls. 1–7.

Nabours, R. K.
 1914. Studies of Inheritance and Evolution in Orthoptera I. *Jour. Genet.*, Vol. 3, No. 3, pp. 141–170.

Pearl, R.
 1917. The Probable Error of a Mendelian Class Frequency. AMERICAN NATURALIST, Vol. 51, No. 603, pp. 144–156.

Platt, F. L.
 1916. "Western Notes and Comment." *Reliable Poultry Journal*, Vol. 23, p. 665.

Punnett, R. C.
 1911. Mendelism. vii + 192 pp. New York: The Macmillan Co.

FURTHER DATA ON THE INHERITANCE OF BLUE IN POULTRY[1]

PROFESSOR WILLIAM A. LIPPINCOTT

KANSAS AGRICULTURAL EXPERIMENT STATION, MANHATTAN, KANSAS

I. PREVIOUS WORK

The principal facts concerning the genetic behavior of blue in the Andalusian breed of domestic fowl were presented in an earlier paper (Lippincott, 1918a). Previous work on the genetics of the blue Andalusian was reviewed and a limited number of further data were offered.

The latter showed that blue Andalusians are like black Andalusians in that they are self-colored. They are, on the other hand, like the blue-splashed Andalusians in that homologous pigmented feathers in both sexes have the same condition with reference to the restriction of pigment in the feather structure. The 1:2:1 ratio obtained from mating blue Andalusians together may be interpreted as the combination of two 3 to 1 ratios. These relationships are shown in Fig. 1.

The restriction of black pigment in the feather structure to give the blue appearance found in blue and in blue-splashed Andalusians was shown to be due to the action of a dominant factor R. The extension of black pigment to all feathers of the body as in both black and blue Andalusians, was found to be due to the action of another dominant factor E.

[1] Contribution from the Department of Genetics, Wisconsin Agricultural Experiment Station, No. 29, and from the Department of Poultry Husbandry, Kansas Agricultural Experiment Station, No. 15.

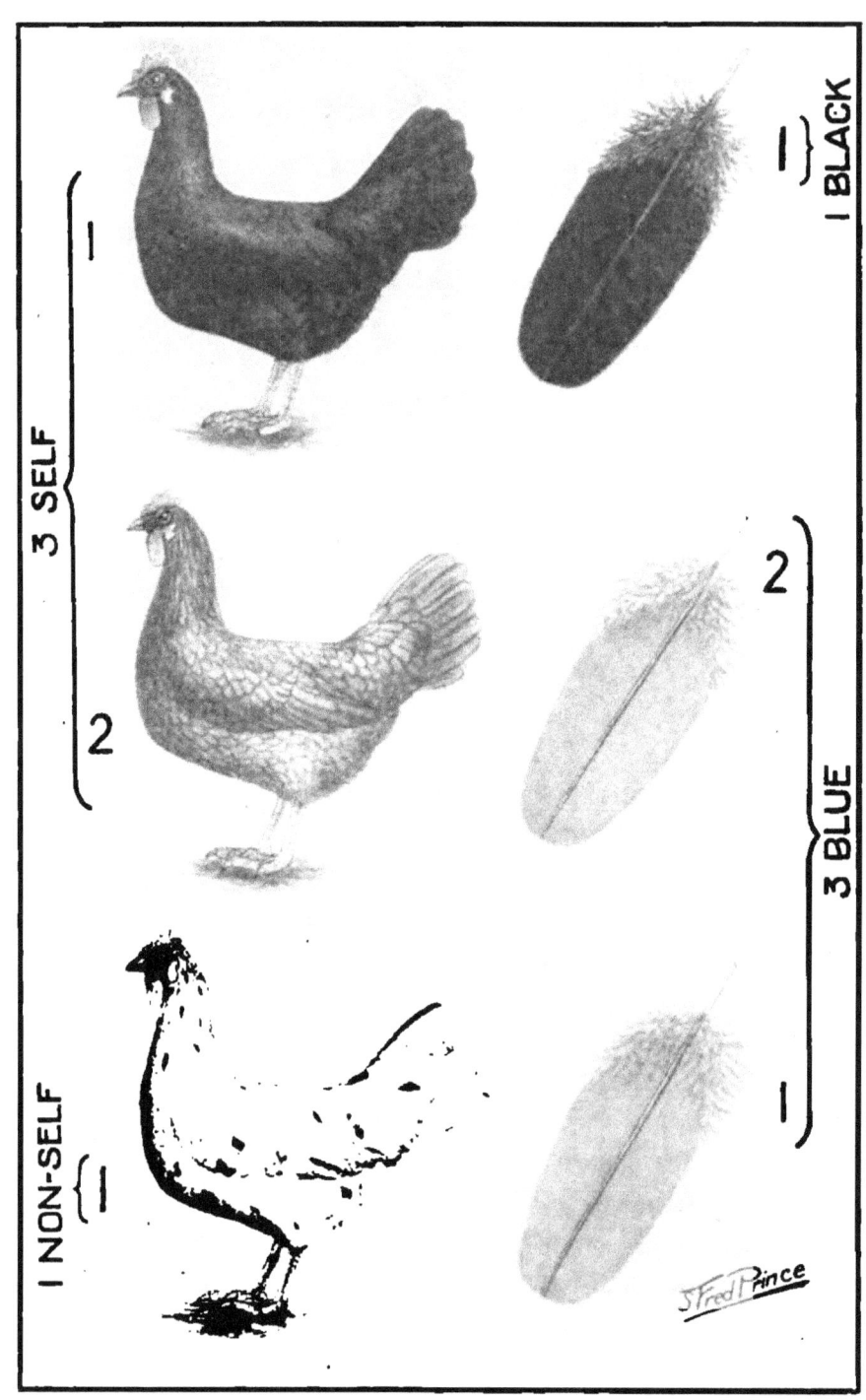

Fig. 1. Showing that the 1:2:1 ratio is a combination of two 3:1 ratios.

It was pointed out that while, on the basis of their expression in the phenotype it appeared more logical to consider these factors as dominants, each closely linked to the recessive allelomorph of the other, they may, so far as the experimental evidence shows, be considered as true allelomorphs occupying identical loci on homologous chromosomes, and each expressing itself independently of the other.

The finding of crossovers between R and E would be conclusive evidence proving the former of the two conditions proposed. It was shown that while no crossovers had been reported the critical data on the case were very limited and the likelihood of crossovers being detected and isolated by breeders is very small. It might well have been suggested further that even though crossing-over does rarely occur, for instance, so that less than one per cent. of the individuals are the product of crossover gametes, the chances of detecting them experimentally are small, considering the limited number of matings (as determined by the equipment available at most experimental institutions) which are likely to be devoted to a search for crossovers.

Though much has been made of the blue Andalusians as a "heterozygote phenotypically intermediate between the parental types" it was shown that while all self-blues so far found had proved to be heterozygous for R and E, they were not in the strict sense intermediate between the parental types. The F_1 progeny of a cross between blue-splashed Andalusians and white Wyandottes was reported as self-blue and far darker than either parent.

It was further shown in the earlier paper that R not only restricts black pigment, so as to render pigmented areas bluish-gray in appearance, but also affects the shape of the pigment granules, so that instead of appearing as rods as in black individuals, they are quite round. In this particular R is quite dominant over its allelomorph, whether one chooses to assume that the latter is E or r.

EXPLANATION OF PLATE I

The photographs shown in Plates I and II were taken by James Machir, my indebtedness to whom it is a pleasure to acknowledge.

FIG. A. Blue Andalusian male.
FIG. B. Blue Orpington female.
FIG. C. Blue-splashed Andalusian female.
FIG. D. Blue Andalusian female.
FIG. E. Blue Orpington male.
FIG. F. Blue-splashed Andalusian male.

A and D—Blue Andalusian.
B and E—Blue Orpington.
C and F—Splashed Andalusian.

It was also shown that both the restricting and the rounding actions of R were interfered with in certain regions of both blue-splashed and blue males. In both color types the pigmented feathers of the neck (hackle), back, and saddle are black or bluish-black instead of blue as on the remainder of the body. The black pigment granules in these regions are for the most part rod-shaped rather than round. It was suggested that this interference with the action of R is a secondary sexual characteristic, presumably due to the presence of testicular or the absence of ovarian influence.

II. Purpose of the Present Paper

It is the purpose of this paper to present further data concerning the inheritance of blue and its relations to the sex glands, and to draw such conclusions as these data justify. A report is given of the breeding behavior of blue as found in the Andalusian, Orpington and Leghorn breeds, and of certain crosses of these breeds with each other and with other breeds, which do not possess blue varieties. The relations of the factors involved to certain factors present in the non-blue varieties of other breeds is considered and evidence concerning the relation of the sex-glands to the action of the factor R presented.

III. Material and Methods

The breeding stock used was from several sources, being in part from the pedigreed flock of the University of Wisconsin, where the work reported in the earlier paper was done. It was also in part from the pedigreed flock of Kansas State Agricultural College where the investigation was continued under the direction of Dr. Leon J. Cole of the University of Wisconsin, my indebtedness to whom it is a pleasure to acknowledge. The stock was, however, mostly from unpedigreed lines, though pure-bred within the meaning of the poultryman. In no case were individuals used which were not from families show-

EXPLANATION OF PLATE II

FIG. A. A black Andalusian male.

FIG. B. A blue F_1 female from a white Wyandotte × blue-splashed Andalusian cross.

FIG. C. A blue (at left) and a black (at right) chick in the down. These are offspring of a white Plymouth Rock ♂ × blue Andalusian ♀. The occipital spots inherited from the sire are plainly visible.

FIG. D. A young blue F_1 male from a blue-splashed × black Langshan cross.

FIG. E. A black Andalusian female.

FIG. F. A blue F_1 male from a white Wyandotte × blue-splashed Andalusian

ing the characteristics of their respective varieties with constancy in so far as could be learned. In as much as only varietal (color), as opposed to breed (shape) characteristics were being studied, less attention was paid to the latter in selecting material. In no case, however, were individuals used which showed disqualifying breed characteristics.

With a single exception no individual was used whose genotype proved to be inconsistent with the "breeding true" of the variety to which it belonged, or, in the case of the blue-splashed Andalusian, the variety from which it arose. This single exception was a blue-splashed Andalusian female (2107) purchased from a breeder who made only blue $\times$ blue matings. She proved to be heterozygous for P, a factor necessary for the production of black pigment. The family from which she arose must have been producing occasional whites which were, in all likelihood, being discarded as extremely light blue-splashed wasters from the blue $\times$ blue matings. This point was not followed up, however, and the facts ascertained. It has been by taking advantage of situations similar to this one that white varieties have been established in several breeds.

There were several individuals discovered whose factorial composition varied from the normal, or usual, for the varieties to which they respectively belonged. Owing to the particular factorial complex of which they were a part, however, these factors behaved as cryptomeres, not affecting the adult phenotype of the variety. Specific reference is made to these individuals in a later section of this paper.

The matings were, for the most part, made in covered

cross. Indications of a factor or factors for lacing may be seen in the hackle and saddle feathers.

FIG. G. A blue-barred (at left) and a black-barred (at right) chick partly feathered. These are offspring of the same mating as the chicks shown in Fig. C this plate. The barring was inherited from their white Plymouth Rock sire.

A and E—Black Andalusian.

B and F—Blue F_1's.

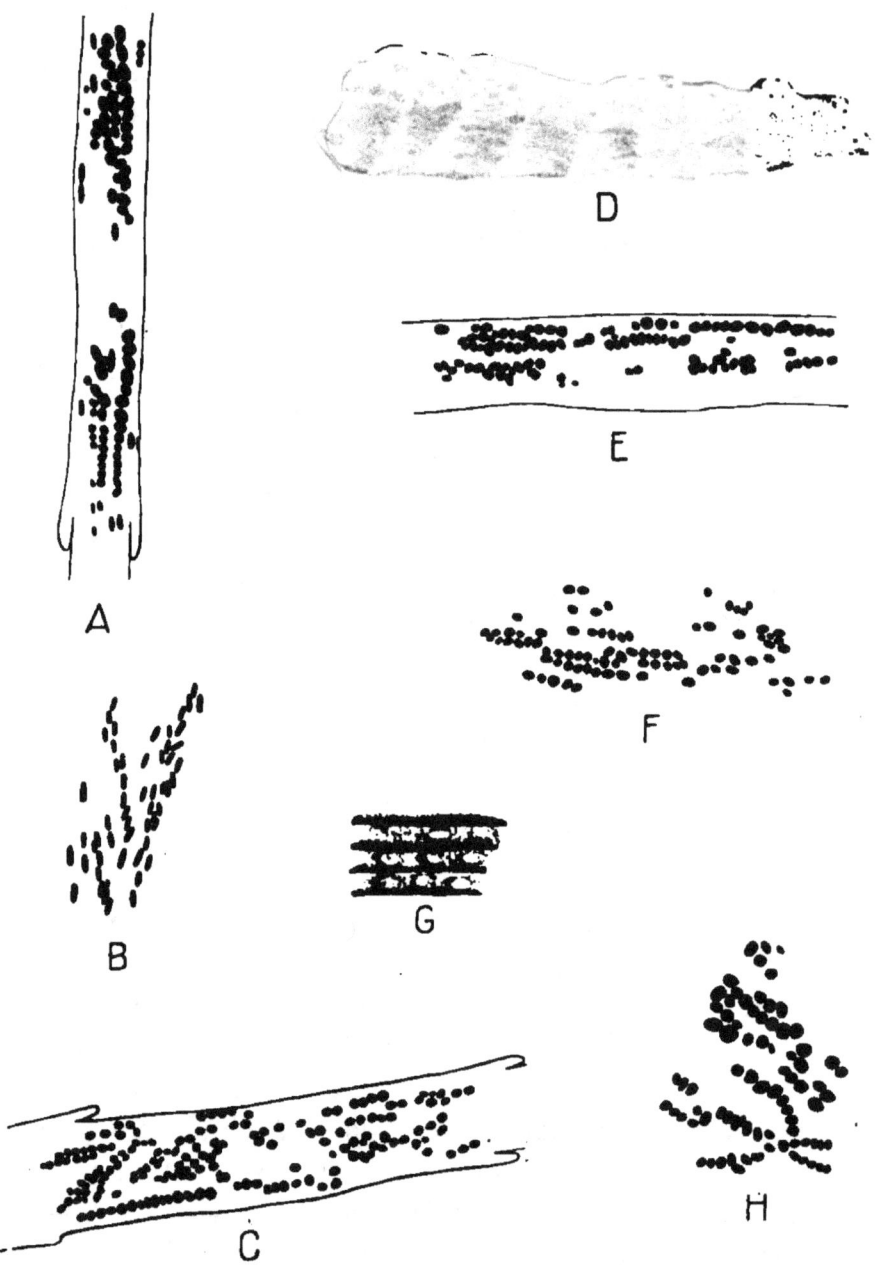

EXPLANATION OF PLATE III

Fig. A. Appearance of pigment granules in a strand of down from a very dark blue crossbred chick. The sire was a blue F_1 from a white Wyandotte × blue-splashed Andalusian cross. The mother was a blue Andalusian. There are occasional rod-shaped granules. Camera lucida drawing.

Fig. B. Appearance of pigment granules in a definitive feather from a black Andalusian. Granules of the same shape are found in the down and definitive feathers of black Langshans, black Orpingtons and the black offspring from crosses with the several breeds used in this investigation. Camera lucida drawing.

Fig. C. Appearance of pigment granules in a strand of down from a blue yards and every precaution taken to insure no mixing of matings and the proper identification of the eggs laid by each female in each mating. Not only was the assistant in charge of the trapnesting selected because of his habitual accuracy in details, but the eggs from each individual hen were kept together, separate from the eggs of other individuals, and carefully compared one with another before being put into the incubator. Any off type or off colored eggs were discarded, so far as these experiments are concerned. In spite of these precautions it is too much to hope that some errors have not crept in, though it is believed they are very few.

Owing to the fact that the original stock was of relatively unknown composition, it was necessary to make such matings as would not only throw light on the behavior of the factors under observation, but would also be likely to bring to light unsuspected factors whose action might intererfere with the action of the genes being studied. This necessitated introducing test females in the matings where the males were uncertain, and of mating many of the females with test males a second season instead of repeating the mating already made. In both cases the result was to reduce greatly the numbers of offspring from some of the crucial types of matings, and considerable numbers of "test" offspring were hatched and described, for the reporting of which here there is no particular object.

The counts of living chicks were made in the down at hatching time and individual descriptions recorded, each

F_1 chick from a white Wyandotte × blue-splashed Andalusian cross. Camera lucida drawing.

FIG. D. Clumped appearance of pigment granules in a curved barbule from a definitive feather of a blue Andalusian.

FIG. E. Appearance of pigment granules in a strand of down from a blue F_1 chick from a blue-splashed Andalusian × white Plymouth Rock cross. Camera lucida drawing.

FIG. F. Appearance of pigment granules in a definitive feather from a blue Audalusian. Camera lucida drawing.

FIG. G. A small area of the web of a definitive feather from a black Andalusian. The cell boundaries and nuclei may be made out. There is no clumping of pigment within the cells.

FIG. H. Appearance of pigment granules in a definitive feather from a blue Andalusian. Camera lucida drawing.

chick being marked with a numbered wingband. The system of keeping pedigree records in use has been described elsewhere (Lippincott, 1918b) and need not be repeated. The descriptions were checked when the chicks were three weeks old, and again at some considerably later, though not specified, time, when the birds were leg-banded for the breeding-pen or laying-house, or were sent to market. The descriptions of all chicks dying were carefully checked at the time they were found, though a small number disappeared without their descriptions being checked. Unless there was reason to suspect that their classification might be likely to change after the taking of the first description such individuals were counted.

Fortunately the different classes of offspring could for the most part be distinguished in the down, and counts were accordingly made of chicks which reached an advanced stage of development but which failed to hatch. It was the practise to test all eggs for live germs at the end of the tenth day of incubation and remove all infertile and dead eggs. A second testing was made on the eighteenth day when all the dead eggs found were opened, the embryos described and their sex recorded. On the twenty-second day, after the hatch was well over, all the eggs which failed to hatch were opened and the descriptions of the dead chicks made a matter of record.

In most cases the embryos from crosses among the three color types of Andalusians which passed the first test, developed far enough so that the differentiation between color types could be made with precision after the down had been carefully washed, and dried with the aid of an electric fan. In those crosses involving recessive white parents, only those unhatched chicks could be counted which lived past the eighteenth day.

There were two possible sources of confusion in the classification of the chicks in the down. These were the differentiations between blacks and occasional very dark blues, and between blue-splashed and recessive whites.

The blacks and dark blues could quite readily be separated by examining the down of each chick microscopically. The blacks carry only rod-shaped pigment granules while in dark blue down rounded granules predominate. These are frequently arranged in rows as reported in my former paper (1918a, pp. 98, 99). In the case of every mating where this method of classification was brought into use to aid in distinguishing individuals which failed to hatch, down samples were saved at hatching time from the living dark blue and black chicks as well, and the first description, the record of the microscopic examination, and the later descriptions after definitive feathers were developed were carefully compared and checked. In nearly all cases the descriptions in the down and in the definitive feathers agreed.

In certain matings, however, it was found that descriptions in the down were not reliable and could not be counted. This was particularly true of one family of Andalusians which carried considerable red in the plumage and which has been referred to by Platt (1916) and Pearl (1917) in other connections. Within this family and its crosses the expression of the R factor in the heterozygote was frequently delayed so that individuals which were described in the down as blacks and which showed only rods under the microscope turned out to be blues when the definitive feathers appeared, and then showed the characteristic round pigment granules of the blue. None of the chicks tracing their ancestry to this family are included in the counts herein reported.

The possible source of confusion in the classification in the down of the blue-splashed and the white chicks arises from the fact that while the adults of the white Plymouth Rocks and white Wyandottes are pure white, or very slightly flecked with black, the chicks frequently carry considerable, though varying amounts of black pigment in the down, which gives certain regions a bluish appearance. This varies in degree from near black to slightly smoky white. Fortunately for the problem in hand the localization of this pigment in the down of certain regions of the body is quite characteristic and quickly recognized. While a blue-splashed chick is frequently very light blue, as noted by Bateson and Punnett (1906, p. 20), the pigment is not localized on the top and back of the head, the wings in the region of the bow, and on the thighs, as it is on the potentially white chick, and the impression conveyed is very different. In potentially white chicks the remiges, which may be seen just starting to grow out from their follicles, are pinkish white and exhibit not the slightest trace of pigment. In the same feathers of the blue-splashed chick, on the other hand, there is a very noticeable bluish cast and usually at least one remex that is distinctly pigmented.

Though in pure-bred white Plymouth Rock and white Wyandotte chicks the pigment granules in the down are typically rod-shaped this fact is not of assistance in classifying with respect to white and blue-splashed offspring from crosses involving the factor R, since under its influence black pigment granules are round whether in a potentially white or a blue-splashed chick.

Not all chicks from pure-bred white Plymouth Rock and white Wyandotte matings exhibit this juvenile pigment. Some can only be recorded as white. It is of interest that the only chicks, three in number, which were originally described as "white, no pigment" or "creamy white" and later used in a breeding pen, have all proved to carry a factor for dominant white, as described in a later section of this paper. The number of such birds which have been tested is small and no general conclusions can be drawn, but the results are suggestive. It is rather interesting to note that a photograph of a group of white Plymouth Rock chicks in "The Plymouth Rock Standard and Breed Book" (American Poultry Association, 1919, p. 419), which is the official guide for the breeding and judging of all Plymouth Rocks, shows individuals which are noticeably pigmented. In response to a letter of inquiry Professor Arthur Smith of the University of Minnesota, the editor of this book, tells me that my observation concerning the presence of pigment in these chicks is correct and he adds in substance that the pigmented chicks develop into the whitest adults.

The fertility and hatching power of the eggs from the various crosses here reported and the viability of the chicks hatched was increasingly disappointing from season to season. While the comparative coefficients of fertility and hatching power have not been calculated, the ratio between the eggs set and chicks hatched has undoubtedly been lower on the average, than for the purebred unrelated matings of the same and other breeds, set in the same incubator at the same time, and certainly lower than would be counted satisfactory in ordinary poultry husbandry practise.

The foregoing applies as well to the rate of mortality. As representative of the numbers surviving to grow definitive feathers in comparison with the counts recorded in the various tables, those of the F_2 from the blue-splashed Andalusian ♂ × white Wyandotte ♀ may be given. The counts made when the chicks were feathered were 47 blue, 18 blue-splashed, 37 black, and 42 white. The total count recorded (see Table IV, group 1) was 100 blue, 46 blue-splashed, 65 black, and 64 white. The reasons for the low hatchability and high mortality have not been established.

Until considerably more data than are now available have been secured it seems best to call attention to the possibility of crossing-over between the loci of R and E by indicating their possible recessive allelomorphs. It is accordingly the practise in this paper to indicate these factors thus: (Re) and (rE).

IV. THE RELATION OF PHENOTYPE TO SEX

It is convenient to consider the relation of phenotype to sex before examining the progenies of the various matings. In order to secure evidence concerning this relation, six blue Andalusian males were caponized during the summer of 1919. Into the body cavities of three of them ovarian tissue from nearly related females was introduced, the other three being kept as checks. The operating was done on July 24 and the birds turned out on range with hundreds of other birds one week later. On September 19 one of them (wingband 1387) was killed by a skunk. At that time it was entirely blue, there being less contrast between the regions that are dark in the male (hackle, back and saddle) and the other regions of the body than frequently appears in blue pullets before comb development indicates the approach of the first laying cycle, and indeed in many mature females. Although it was over four months old (hatching date, May 6, 1919) it appeared so much like an immature pullet that it was mistaken for one by the poultryman in charge and by the writer, until its record and description were consulted. Concerning the latter there did not seem to be any chance of error, since the scar made in opening the body cavity was plainly visible.

Such a situation indicates a fairly complete molt between July 24 and September 19. This is not surprising, however, since Rice, Nixon and Rogers (1908, p. 66) have shown that "from the incubator to the laying period the chicks experienced at least four molts, either partial or complete," and it is further well known that a close relation exists between molting and ovarian activity.

The other birds operated on at the same time were at once looked up and described. One of them (wingband 1855) was found to be somewhat intermediate in condition, some of the feathers of the neck and saddle being blue, but somewhat darker in shade than the normally (in the male) blue regions of the body. There were, however, a few scattered feathers which were almost black from the tip halfway down the web toward the fluff. About midway between the tip of the feather and the beginning of the fluff there was a distinct line of demarcation where the black or near-black became a distinct blue. This chick was hatched a little over two weeks later (May 22) than 1387 and had apparently not gone through a complete molt, some feathers in process of growth at the time of the operation and showing ovarian influence on the last regions to develop still remaining.

On October 26 this bird was killed, apparently by a rat, at which time all of the feathers of the neck and saddle regions were distinctly blue, though considerably darker than other parts of the body. The shape of the feathers was characteristically female.

The third male into which ovarian tissue was introduced (wingband 1480) showed no influence of the introduced tissue on September 19. This condition still prevailed when it was sent to market October 26. It appeared normal for a blue capon of that age, over five months, the hackle and saddle being very dark and characteristically male in shape. Presumably the ovarian tissue introduced atrophied without having any effect.

Of the cockerels which were caponized, but had no ovarian tissue introduced, one (wingband 1859) died soon after the operation. The other two (wingbands 1415 and 1492) showed and continued to show typical blue capon characteristics with regard to the color and shape of the saddle and hackle feathers. The feathers were fully as dark as in normal males of the same age, and as they matured were even longer than their homologs in normal males. This result is precisely the same as that observed by the writer several times in blue capons, concerning which no descriptive records were kept.

In this connection it should be observed that in the family of Andalusians here dealt with, it has been not infrequently noticed that certain nearly grown pullets whose combs have not begun to develop, show only very dark feathers in the regions of the neck and back. These same birds after their combs begin to redden, thereby indicating ovarian activity and the onset of laying, appear to pass through a molt or partial molt whereby the dark feathers of the back region particularly, are gradually replaced by those of a clearer blue. The necks of such females usually remain dark, showing considerable contrast with the other regions of the body, though being by no means as dark as the same region of the blue male.

Although the number of desexed males into which ovaries were introduced was small, it seems fair to conclude in the light of the evidence concerning testicular (Goodale, 1916) and ovarian (Goodale, 1918; Cole and Lippincott, 1919) influence in fowls that the failure of the factor R to express itself as fully in the neck, back and saddle regions of the blue and blue-splashed males as in the females is due to the lack of some necessary cooperative action on the part of the ovary, and not to any inhibitive action on the part of the testis.

V. The Breeding Behavior of Andalusians

New data concerning the breeding behavior of the three color types of Andalusians, as shown by several types of matings, are presented in Table I.

TABLE I
Showing the Numbers and Color Types of Progenies from Various Andalusian Crosses[2]

Group	♂♂	♀♀		Blue-spl. $(Re)(Re)$	Blue $(Re)(rE)$	Black $(rE)(rE)$
1..	Blue $(Re)(rE)$	× blue $(Re)(rE)$	Obtained... Theoretical.	46 53.5	104 107	64 53.5
2..	Blue $(Re)(rE)$	× black $(rE)(rE)$	Obtained... Theoretical.	00 00	25 24.5	24 24.5
3..	Black $(rE)(rE)$	× blue $(Re)(rE)$	Obtained... Theoretical.	00 00	113 101.5	90 101.5
4..	Blue $(Re)(rE)$	× blue-splashed $(Re)(Re)$	Obtained... Theoretical.	1 1	1 1	0 0
5..	Blue-splashed $(Re)(Re)$	× blue $(Re)(rE)$	Obtained... Theoretical.	35 34	33 34	0 0
6..	Black $(rE)(rE)$	× blue-splashed $(Re)(Re)$	Obtained... Theoretical.	0 0	138 138	0 0
7..	Blue-splashed $(Re)(Re)$	× black $(rE)(rE)$	Obtained... Theoretical.	0 0	56 56	0 0
8..	Blue-splashed $(Re)(Re)$	× blue-splashed $(Re)(Re)$	Obtained... Theoretical.	0 0	12 12	0 0

[2] Andalusians are normally homozygous for P, a factor necessary for the production of black pigment.

These results are in substantial accord with those of Bateson and Punnett (1906, p. 20). A somewhat marked departure from the theoretical expectation appears in group (3) of black ♂♂ × blue ♀♀ matings, the agreement in the reciprocal cross (group 2) being as close as possible. This departure from expectation is due to the progeny of a single pair of birds (♂136M and ♀2005) which produced 25 blues and 6 blacks. If the latter are left out of consideration the results are 88 blues and 84 blacks.

However, even in the case of the progeny of ♂136M and ♀2005 the Dev./P.E. = 4.1, which indicates a deviation of doubtful significance. The results of this mating were carefully considered from the standpoint of crossing-over, but there is no indication of its having occurred.

According to these results the genetic compositions of the three color types of Andalusians used in these experiments were as follows: blue-splashed = $(Re)(Re)$, blue = $(Re)(rE)$, and black = $(rE)(rE)$. There was no evidence of crossing-over between R and E having occurred.

VI. Data from Crosses of Andalusians with Certain Recessive White Breeds

In the previous paper (1918a, p. 106) the writer reported a small number of data on a cross between a white Wyandotte ♂ and a blue-splashed Andalusian ♀. These have been considerably increased in amount and the reciprocal cross made. Further, both blue and black Andalusians have been crossed reciprocally with white Wyandottes and all three Andalusian color types crossed reciprocally with white Plymouth Rocks. The data from these several matings are set forth in Table II.

The crosses were made in the twelve possible ways, from eleven of which offspring were secured, the one type of mating which failed to produce offspring being the white Wyandotte ♂ × black Andalusian ♀. Inasmuch as there is no evidence that any of the factors here under observation are sex-linked and there is considerable evidence that they are not, this omission is not serious.

TABLE II

SHOWING THE RESULTS OF CROSSING THE THREE-COLOR TYPES OF ANDALUSIANS WITH WHITE WYANDOTTES AND WHITE PLYMOUTH ROCKS

Group	♂♂	♀♀		Blue	Black
1.	Blue-splashed Andalusian $PP(Re)(Re)$	× white Wyandotte $pp(rE)(rE)$	Obtained Theoretical	65 65	00 00
2.	White Wyandotte $pp(rE)(rE)$	× blue-splashed Andalusian $PP(Re)(Re)$	Obtained Theoretical	50 50	00 00
3.	Blue-splashed Andalusian $PP(Re)(Re)$	× white Plymouth Rock $pp(rE)(rE)$	Obtained Theoretical	179 179	00 00
4.	White Plymouth Rock $pp(rE)(rE)$	× blue-splashed Andalusian $PP(Re)(Re)$	Obtained Theoretical	87 87	00 00
5.	Blue Andalusian $PP(Re)(rE)$	× white Wyandotte $pp(rE)(rE)$	Obtained Theoretical	27 25.5	24 25.5
6.	White Wyandotte $pp(rE)(rE)$	× blue Andalusian $PP(Re)(rE)$	Obtained Theoretical	13 15.5	18 15.5
7.	Blue Andalusian $PP(Re)(rE)$	× white Plymouth Rock $pp(rE)(rE)$	Obtained Theoretical	80 67.5	55 67.5
8.	White Plymouth Rock $pp(rE)(rE)$	× blue Andalusian $PP(Re)(rE)$	Obtained Theoretical	24 28	32 28
9.	Black Andalusian $PP(rE)(rE)$	× white Wyandotte $pp(rE)(rE)$	Obtained Theoretical	00 00	18 18
10.	Black Andalusian $PP(rE)(rE)$	× white Plymouth Rock $pp(rE)(rE)$	Obtained Theoretical	00 00	132 132
11.	White Plymouth Rock $pp(rE)(rE)$	× black Andalusian $PP(rE)(rE)$	Obtained Theoretical	00 00	28 28

The results of these crosses are understandable on the assumption suggested in the earlier paper that the individuals from the recessive white races are homozygous for the factors E and p, p being the recessive allelomorph of P, a factor necessary for the production of black pigment in the feathers. Sturtevant (1912) first suggested that Wyandotte white is recessive, a fact which was overlooked in my earlier paper. Morgan and Goodale (1912, p. 115) have made a similar assumption for the white Plymouth Rock.

Since in the series of experiments being reported here, reciprocal crosses of white Wyandottes and white Plymouth Rocks gave only whites, thereby showing no evidence of recombination, it seems fair to assume that the white of both breeds is due to the same recessive factor p in homozygous condition.

The condition of the white Rocks and white Wyandottes reported in Table II, with reference to E, appears clear, since in all crosses with blue-splashed Andalusians (and as will appear later, in the case of the Wyandotte, with blue-splashed Orpingtons) which are homozygous for P and R, but do not carry E, all offspring, 381 in number, were without an exception, blue (see mating groups 1 to 4, Table II).

On this basis blue Andalusians, $PP(Re)(rE)$, mated with such recessive whites should produce blues and blacks in equal numbers. Mating groups 5 to 8, inclusive, in Table II show the results of such matings, which combined give 144 blues to 129 black (136.5 to 136.5 would be equality), a fair realization of the expectation.

As would be expected from the foregoing, crosses of similar recessive whites with black Andalusians ($PP(rE)(rE)$) (see Table II, groups 9 to 11, inclusive) gave only blacks. Of these there were in all 178 individuals and no exceptions.

The offspring of the crosses reported in Table II frequently gave evidence that the recessive white parents carried pattern factors as cryptomeres, but for the sake of clearness these complications, which have nothing directly to do with the study in hand, have been ignored in summarizing the data. As was to be expected, the white Plymouth Rocks carried the sex-linked pattern factor for barring. All pigmented offspring by a white Rock sire showed evidences of barring as soon as the definitive feathers appeared. Two such, the offspring of a white Plymouth Rock ♂ and blue Andalusian ♀ are shown in Fig. G, Plate II. Even at hatching, the occipital spot, which may be a juvenile effect of the factor for barring, gave notice of the presence of the barring factor. In the work here reported it was found possible to classify in the down pigmented offspring of a non-barred ♂ × white Plymouth Rock ♀ cross accurately with regard to sex, by the presence or absence of the occipital spot. Morgan and Goodale (1912) made use of this spot in classifying barred and non-barred chicks which failed to hatch and Punnett (1919) also has made use of it in sorting newly hatched cross-bred chicks according to sex.

The progeny of crosses involving white Wyandottes frequently displayed Wyandotte lacing of a lesser or greater degree of perfection, though the appearance of this pattern was neither as constant nor as distinct as that of the barred pattern. The appearance of the lacing was to be expected if, as is generally stated in the literature on Wyandottes (see McGrew, 1901), the white variety was derived directly from the silver Wyandotte, which is laced.

In connection with these recessive white crosses is to be noted the fact that several white individuals, although "pure-bred" in the terminology of the poultryman, gave results which differed from the foregoing. Four white Wyandotte females proved to carry both the R and E factors and were of the same composition with respect to these factors as a pure-bred blue Andalusian, but unlike the blue Andalusian they carried p in the homozygous condition. One of these, which has already been reported on elsewhere (Lippincott, 1919), carried the sex-linked pattern factor for barring as well. Dryden (1916, p. 67) has also reported a white Wyandotte carrying a factor for barring.

One white Plymouth Rock and eight white Wyandottes proved to be heterozygous for a factor for dominant white. These were tested and found to be homozygous for p. In other words they carried both dominant and recessive white. Bateson and Punnett (1905, p. 117) appear to have had birds of this type and Dryden (1916, p. 66) reports a white Wyandotte which produced only white chicks when mated to a black Minorca, hence must have been homozygous for a dominant white factor. Whether it carried P or p, the evidence does not show.

So far no attempt has been made to ascertain whether this factor for dominant white is the same as that normally carried by the white Leghorn and which Hadley (1913 and 1914) designated as I. For convenience and to recognize the possibility of its differing from I the factor here dealt with is referred to in this paper as I^P (inhibitor of pigment) and its allelomorph as i^P.

VII. Back-crosses of F_1's from Blue-splashed Andalusian × Recessive White Matings

The results of crossing the F_1 blues from the blue-splashed Andalusian × recessive white crosses is shown in Table III.

While by no means all possible back-crosses have been made, enough are represented to show clearly that factors R and E were appearing in approximately equal numbers, and that this was also true of P and p, though in some cases the presence of I^P complicated matters somewhat. It was, unfortunately, not always possible to use the actual parents in making back-crosses and though individuals from the same families were employed, this proved to be no criterion that they would be of the same genotype as the individuals used in the original cross. There can be no question as to their factorial composition, however, as each individual has been either deliberately tested or had happened to be so mated for another purpose as to give dependable evidence on its composition with respect to I^P and p.

So far as it goes, the evidence, which is substantiated by the results of other crosses to be reported in a later section of this paper, also shows that the meeting of P and R was according to chance, thereby indicating no linkage between these two factors.

It will be noted that the blue F_1 ♀♀ in group 5 of Table III had a blue Andalusian mother instead of a blue-

TABLE III

Showing the Results of Back-crossing F_1 Blues from Blue and Blue-splashed Andalusian × Recessive White Crosses

Group	♂♂	♀♀		Blue-splashed	Blue	Black	White
1....	Blue F_1 $\frac{\text{spl. And. ♂}[a]}{\text{wh. Wyand. ♀}}$ $Pp(Re)(rE)$	× blue-splashed Andalusian $PP(Re)(Re)$	Obtained Theoretical	15 13.5	12 13.5	00 00	00 00
2....	Blue F_1 $\frac{\text{wh. Wyand. ♂}}{\text{spl. And. ♀}}$ $Pp(Re)(rE)$	× blue-splashed Andalusian $PP(Re)(Re)$	Obtained Theoretical	4 3.5	3 3.5	00 00	00 00
3....	Blue F_1 $\frac{\text{spl. And. ♂}}{\text{wh. Rock ♀}}$ $Pp(Re)(rE)$	× blue-splashed Andalusian $PP(Re)(Re)$	Obtained Theoretical	25 24.5	24 24.5	00 00	00 00
4....	Blue F_1 $\frac{\text{wh. Wyand. ♂}}{\text{spl. And. ♀}}$ $PpiP_iP(rE)(rE)$	× white Wyandotte $ppIP_iP(rE)(rE)$	Obtained Theoretical	00 00	12 10.375	7 10.375	04 62.250
5....	White Plymouth Rock $pp(rE)(rE)$	× blue F_1 $\frac{\text{blue And. ♀}}{\text{wh. Wyand. ♂}}$ $Pp(Re)(rE)$	Obtained Theoretical	00 00	8 10	8 10	24 20
6....	White Wyandotte $ppiP_iP(rE)(rE)$	× blue F_1 $\frac{\text{spl. And. ♂}}{\text{wh. Wyand. ♀}}$ $Ppi^iP^iP(Re)(rE)$	Obtained Theoretical	00 00	3 5.375	7 5.375	33 31.250

[a] This convention in this and subsequent tables is used to indicate the kind and direction of the original cross.

splashed. From the nature of the behavior of the factors R and E already described, this would make no difference with regard to the blue offspring, for the blue progeny of a blue Andalusian female by a white Wyandotte male would be of exactly the same composition with respect to R, E, and P as *all* the offspring of a blue-splashed Andalusian mother by the same sire.

It will also be noted in this group (5) that while the father of the F_1 blue was a white Wyandotte, the male used in this cross was a white Plymouth Rock. Since it has been shown that for the factors being studied, white Plymouth Rocks and white Wyandottes are identical, this should not affect the ratios.

VIII. The F_2 Ratios from Blue-splashed Andalusian × Recessive White Matings

The F_2 ratios from various blue-splashed Andalusian × recessive white crosses are shown in Table IV.

As will be seen, the four F_2 classes predicted for such crosses in the writer's earlier paper (1918a, p. 113) on the basis of the F_1 results, have been obtained. No other classes have appeared. This would seem to indicate that the factorial compositions of the blue-splashed Andalusians and white Wyandottes then proposed were correct and that the white Plymouth Rocks used were of the same composition with respect to the factors R, E and P as were the white Wyandottes.

Seven F_1 blue males were used in securing the F_2 ratios. The legband numbers of these males may be found in Table IV, in the column headed "Band No." The direction of the original cross is indicated for each male and for the group of females with which he was mated. The direction of the cross was the same for the males and the females in all cases but two. Males 296M and 258M were mated with females which were products of the same crosses, respectively, as they themselves (groups 2 and 7), and also with females from the reciprocal crosses (groups 3 and 8).

TABLE IV

Showing F_2 Ratios from Crosses of Blue-splashed Andalusians and White Wyandottes and White Plymouth Rocks[a]

Group	♂	Band No. ♀♀	♀		Blue	Blue-spl.	Black	White
1...	Blue F_1 spl. And. ♂ / wh. Wyand. ♀ $X^2 = 4.5747$	86E P = .2099	× blue F_1 spl. And. ♂ / wh. Wyand. ♀	Obtained Theoretical	100 103.1250	46 51.5625	65 51.5625	64 68.7500
2...	Blue F_1 spl. And. ♂ / wh. Wyand. ♀ $X^2 = 4.0998$	296M P = .252516	× blue F_1 spl. And. ♂ / wh. Wyand. ♀	Obtained Theoretical	73 69	24 34.5	39 34.5	48 46
3...	Blue F_1 spl. And. ♂ / wh. Wyand. ♀ $X^2 = .8759$	296M P = .83592	× blue F_1 wh. Wyand. ♂ / spl. And. ♀	Obtained Theoretical	45 40.875	17 20.4375	20 20.4375	27 27.2500
4...	Blue F_1 wh. Wyand. ♂ / spl. And. ♀ $X^2 = 5.7301$	66M P = .1278	× blue F_1 wh. Wyand. ♂ / spl. And. ♀	Obtained Theoretical	19 17.25	14 8.625	7 8.625	6 11.5
5...	Blue F_1 wh. Wyand. ♂ / spl. And. ♀ $X^2 = 4.6871$	65E P = .1998	× blue F_1 spl. And. ♂ / wh. Wyand. ♀	Obtained Theoretical	94 82.875	41 41.4375	44 41.4375	42 55.2500
6...	Blue F_1 spl. And. ♂ / wh. Rock ♀ $X^2 = 2.7349$	40E P = .4395	× blue F_1 spl. And. ♂ / wh. Rock ♀	Obtained Theoretical	34 29.625	10 14.8125	17 14.8125	18 19.75
7...	Blue F_1 spl. And. ♂ / wh. Rock ♀ $X^2 = 6.7510$	258M P = .081786	× blue F_1 spl. And. ♂ / wh. Rock ♀	Obtained Theoretical	51 62.625	26 31.3125	39 31.3125	51 41.75
8...	Blue F_1 spl. And. ♂ / wh. Rock ♀ $X^2 = 8.20$	258M P = .042668	× blue F_1 wh. Rock ♂ / spl. And. ♀	Obtained Theoretical	14 7.8750	3 3.9375	3 3.9375	1 5.25
	Total ratios for all crosses $X^2 = 8.6410$ P = .0353			Obtained Theoretical	430 413.25	181 206.625	234 206.625	257 275.500

[a] The formulæ of color types involved in these crosses are: blue-splashed Andalusian PP (Re) (rE), white Rocks and Wyandottes $pp(rE)(rE)$, and F_1 blues $Pp(Re)(rE)$.

As may be seen by inspection of Table IV, but one male (296M) gave a group of offspring (3) which was very close to expectation. The chances that as great a deviation as this one would appear as a result of random sampling are four to one. The mothers of this group were the product of a cross which was the reciprocal of that which produced their sire. The offspring of 296M when mated with females which were the product of the same cross as himself (group 2) gave a deviation so great that the chances against its appearing as a result of random sampling are three to one. The chances of the appearance of deviations as great as those shown by the offspring groups of the other males were as follows: 86E (group 1) one chance in a little less than five; 66M (group 4) one chance in about eight; 65E (group 5) one chance in approximately five; 46E (group 6) one chance in about two and a quarter; 258M (group 7) once in about twelve times with females from the same cross as he, and once in twenty-five when mated with females from a reciprocal cross (group 8).

It would be unusual, though not impossible, to have so many comparatively wide deviations from expectation simply as a result of random sampling.

If the genetic constitution of the F_1's was as has been previously postulated, and these were in fact all chance deviations, it would be highly probable that the lumping of all the data given in Table IV would approximate the calculated ratio fairly closely.

The lumped data are given at the bottom of Table IV. It will at once be seen that the goodness of fit as measured by P is poorer than the poorest constituent group, and would be probable, on the basis of random sampling, once in about twenty-eight times. It seems fairly clear that some disturbing force was operative.

The two possible causes of disturbance which present themselves are linkage and a differential viability of classes, or it might be a combination of the two.

Linkage between the two principal pairs of factors involved in the crosses, Pp and $(Re)(rE)$, may not be appealed to because the only possible linkage relation would produce results diametrically opposed to those with which we are confronted. Since according to our hypothesis the recessive white parents were in each case of the composition $pp(rE)(rE)$ and the blue-splashed Andalusian parent $PP(Re)(Re)$, it is evident that linkage would require the production of $p(rE)$ gametes by the F_1 blues, more often than $P(rE)$ gametes. And similarly the combination $P(Re)$ should also appear more often than $p(Re)$.

A complete linkage between these pairs of allelomorphs would result in an F_2 ratio of 1 blue-splashed and 2 blue to 1 white, the blacks not appearing. The tendency of even weak linkage would be to reduce the proportional number of blacks. This should be true irrespective of the direction of the cross. It would further be true, that unless crossing-over occurred in both sexes any linkage whatsoever would inhibit the production of F_2 blacks homozygous for P. As will be shown in a later section of this paper, however, F_2 blacks homozygous for P have been identified. Even a casual inspection of Table IV shows that a relative preponderance of blacks is a quite constant characteristic.

Crossing-over in the male fowl has been found by Goodale (1917) and in the male pigeon by Cole and Kelley (1919). The latter investigators definitely state that there is no crossing-over of sex-linked factors in the female pigeon. Goodale states that none had been observed in the female fowl, but that a definite test of the matter would be made later. So far as the writer is aware no further report has been made. It should perhaps be pointed out that so far only sex-linked factors have been dealt with, no autosomal linked groups in birds having so far been reported.

There are no F_2 data available from crosses where p and (Re) are found in one parent and P and (rE) in the other. The F_1's from such a cross have been secured by mating an extracted white of the composition $pp(Re)(Re)$ with a black Andalusian, $PP(rE)(rE)$, which gave all blues. From these an attempt will be made to secure F_2's in considerable numbers. Back-crosses to the parental types will also be made. The F_2's should approximate the same ratios as appear in Table IV and also give some evidence on the second possible explanation of the persistent deviations about to be discussed.

The calculation of theoretical expectancies presupposes the equal viability of all phenotypic and genotypic classes. If for any reason the individuals of one or more of the obtained classes tend to be less viable than certain other classes, deviation from expectancy will occur if the lack of viability expresses itself prior to making the counts.

As has already been pointed out, the lumping of the data presented in Table IV brings forth a poorer fit than is shown in any of the constituent groups. The deficient classes are the blue-splashed and the white, while the most preponderant class relatively is the black.

It seems to be a rather tacit assumption among poultrymen, particularly, it must in truth be said, among those breeding pigmented varieties, that the recessive white varieties are less vigorous (and so in all probability less viable) than the pigmented varieties of the same breeds. In how far this assumption is based on fact there is no critical evidence to call upon.

Regarding the relative viability of splashed and self-colored races there is no suggestion from any source. Splashed varieties are, so far as I am aware, nowhere bred as such, and the experience of practical breeders may accordingly not be appealed to.

While in the case in hand the assumption of low viability on the part of the individuals of the splashed and recessive white classes seems to correspond with the facts, such an assumption, though convenient, is not corroborated by other evidence. That the splashed classes are not necessarily always deficient is shown by the progeny of the blue-splashed × blue mating in Table I, group 5, and of the F_1 blue × blue-splashed matings in Table III, groups 1, 2 and 3.

The latter fact suggests that possibly certain individuals used in these matings carried recessive factors tending to cause low viability, which were linked to the factor R. Until the fact of a differential viability is demonstrated, however, it is useless to speculate on this possibility. The reason for the deficiencies in the blue-splashed and also in the white classes, therefore can not at present be determined.

IX. Identification of the F_2 Genotypes

As indicated in my former paper (1918a, p. 113) the genotypes expected in the several F_2 phenotypes from the blue-splashed × recessive white crosses are as follows: blue, $PP(Re)(rE)$ and $Pp(Re)(rE)$; blue-splashed, $PP(Re)(Re)$ and $Pp(Re)(Re)$; black, $PP(rE)(rE)$ and $Pp(rE)(rE)$; white, $pp(Re)(Re)$, $pp(Re)(rE)$ and $pp(rE)(rE)$. Although the limitations of equipment were such that comparatively few F_2 individuals could be tested, fortunately all of the genotypes but one have been identified by making the appropriate crosses. The blues mated to individuals homozygous for p and E gave blues and blacks in equal numbers, or, blues, blacks and whites in the approximate ratio of $1:1:2$, as the case might be. The blue-splashed mated to individuals of the same constitution produced all blues, or, equal numbers of blues and whites, depending upon whether or not they were homozygous with respect to P. Similarly the blacks gave all blacks, or, blacks and whites, depending upon their condition with respect to P.

The whites on the other hand were mated to blacks known to be homozygous for P and E. The $pp(Re)(Re)$ whites, as mentioned in an earlier section of this paper, gave all blues, just as would blue-splashed Andalusians. The $pp(Rc)(rE)$ whites produced blacks and blues in approximately equal numbers, exactly as would blue Andalusians. The parental white genotype $pp(rE)(rE)$, which would give all blacks, was curiously enough, the one of the whites which did not happen to be selected for testing.

It is important to note that while eight out of the nine F_2 genotypes were identified, no genotypes were found other than those expected.

X. Data on Andalusian × Black Langshan Crosses

It appeared desirable, in order to ascertain whether there was anything inherent in Andalusian black which made its relation to Andalusian blue different from that of other black breeds, to make certain matings of Andalusians with black Langshans. The Langshan was chosen because not only is it a different breed, but it also belongs to a different group of breeds. The original black Langshans were, according to Brown (1906, p. 63), imported from China, while the Andalusians, according to the same authority (p. 107), originated from native stocks along the borders of the Mediterranean Sea. So far as is known they have nothing in common in their immediate ancestry. Davenport (1914) even points to the probability that the immediate wild ancestors of the Asiatic breeds differed from those of the Mediterranean breeds. If blacks differ in their relation to Andalusian blue it would seem probable that Andalusian black and Langshan black might show this difference.

The results of the Andalusian-Langshan matings are shown in Table V. As may be seen readily by reference to this table the results are in every case precisely those which might be expected if a black Andalusian had been substituted for the black Langshan. So far as the principal factors under discussion are concerned it appears that the black Langshans used were identical in composi-

TABLE V

Showing the Results of Several Andalusian × Black Langshan Crosses

Group	♂	♀♀		Blue Splashed	Blue	Black
1	Blue Andalusian $PP(Re)(rE)$	× black Langshan $PP(rE)(rE)$	Obtained Theoretical	0 0	34 32.50	31 32.50
2	Blue Andalusian $PP(Re)(rE)$	× black wh. Wyand. ♂ black Lang. ♀ $Pp(rE)(rE)$	Obtained Theoretical	0 0	15 18	21 18
3	Blue-splashed Andalusian $PP(Re)(Re)$	× black Langshan $PP(rE)(rE)$	Obtained Theoretical	0 0	65 65	0 0
4	Black Andalusian $PP(rE)(rE)$	× black Langshan $PP(rE)(rE)$	Obtained Theoretical	0 0	0 0	11 11
5	Blue $\dfrac{\text{blue And. ♂}}{\text{black Lang. ♀}}$ $PP(Re)(rE)$	× blue $\dfrac{\text{blue And. ♂}}{\text{black Lang. ♀}}$ $PP(Re)(rE)$	Obtained Theoretical	4 4.5	9 9	5 4.5
6	Blue $\dfrac{\text{blue And. ♂}}{\text{black Lang. ♀}}$ $PP(Re)(rE)$	× black $\dfrac{\text{blue And. ♂}}{\text{black Lang. ♀}}$ $PP(rE)(rE)$	Obtained Theoretical	0 0	12 10.5	9 10.5
7	Blue $\dfrac{\text{blue And. ♂}}{\text{black Lang. ♀}}$ $PP(Re)(rE)$	× black Langshan $PP(rE)(rE)$	Obtained Theoretical	0 0	12 12.5	13 12.5
8	Blue $\dfrac{\text{blue And. ♂}}{\text{black Lang. ♀}}$ $PP(Re)(rE)$	× blue Andalusian $PP(Re)(rE)$	Obtained Theoretical	5 3.25	6 6.5	2 3.25

tion with the black Andalusians, being $PP(rE)(rE)$. The condition of the Langshan with respect to P was found by mating individuals with white Wyandottes, whereby only black, i.e., pigmented, offspring were produced.

XI. THE RELATION OF ORPINGTON BLUE TO ANDALUSIAN BLUE

Among the Orpingtons, an English breed, is a blue variety. Like the blue Andalusian it is an inconstant breeder with regard to color, segregating into blue-splashed and blacks as well as blues. Though by no means as widely bred as the blue Andalusians, it has numerous admirers, some of whom have claimed verbally to the writer that the proportion of wasters, i.e., blue-splashed and blacks, was much smaller than in the Andalusians, though no figures are obtainable by way of sub-

TABLE VI

Showing the Results of Certain Blue, Blue-splashed and Black Orpington Crosses among Themselves and with Other Breeds

Group	♂	♀♀		Blue-spl.	Blue	Black	White
1...	Blue Andalusian $(Re)(rE)$	× blue Orpington $(Re)(rE)$	Obtained	13	30	12	0
			Theoretical	13.75	27.50	13.75	0
2...	Blue-splashed Andalusian $(Re)(Re)$	× black Orpington $(rE)(rE)$	Obtained	0	37	0	0
			Theoretical	0	37	0	0
3...	Black Andalusian $(rE)(rE)$	× blue-splashed Orpington $(Re)(Re)$	Obtained	0	14	0	0
			Theoretical	0	14	0	0
4...	White Wyandotte $pp(rE)(rE)$	× blue-splashed Orpington $PP(Re)(Re)$	Obtained	0	21	0	0
			Theoretical	0	21	0	0
5...	Blue Orpington $(Re)(rE)$	× blue Orpington $(Re)(rE)$	Obtained	19	58	15	0
			Theoretical	23	46	23	0
6...	Blue F_1 wh. Wyand. ♂ spl. Orp. ♀ $Pp(Re)(rE)$	× blue F_1 wh. Wyand. ♂ spl. Orp. ♀ $Pp(Re)(rE)$	Obtained	60	26	24	25
			Theoretical	50.6250	25.3125	25.3125	33.7500

$X^2 = 4.0505$ $P = .2569$

stantiation. It seemed desirable from several standpoints to ascertain what factors were involved in the production of Orpington blue, and whether the blue Orpington differed from the blue Andalusian in its genetic behavior. A number of matings were accordingly made, the data from which are shown in Table VI.

These data are consistent with the supposition that the factors involved in the production of Orpington blue are identical with those which produce Andalusian blue. The crossing of blue Andalusians and blue Orpingtons gave exactly the same sort of result as that obtained by mating blue Andalusians *inter se,* as shown by group 1. The blue-splashed Orpingtons mated with white Wyandottes gave only blues (group 4) just as did the blue-splashed Andalusians. And finally the F_2 ratio from white Wyandotte × blue-splashed Orpington crosses gave the same phenotypic classes as were obtained in the F_2 from the white Wyandotte × blue-splashed Andalusian cross, with a deviation from expectancy as great as would be probable once in four times. It is interesting to note that while the white class is deficient in this case, the blue-splashed class is not.

XII. Data from Blue Leghorn Crosses

In the spring of 1917 there appeared in the large pure-bred single comb white Leghorn flock of the Pabst Stock Farm at Oconomowoc, Wisconsin, two blue females. The flock was not pedigreed and nothing is known of the individual ancestors of these birds. They were of fair Leghorn type and were, as far as known, the offspring of pure-bred white Leghorn parents. Through the courtesy of Mr. Fred Pabst, and Dr. L. J. Cole of the University of Wisconsin, these individuals came into the hands of the writer and were entered on the records of the Department of Poultry Husbandry of the Kansas State Agricultural College as numbers 767 and 768.

Number 767 was a fairly even shade of medium to light

blue when received and showed some evidence of barring, though this was not very distinct. Number 768 was much lighter in shade than 767 and showed no evidence of barring. In contrast with ordinary blue she would, from a little distance, be mistaken for a white. The pigment granules in both cases were round.

The results of mating these birds in various ways are presented in Table VII. The numbers are rather small

TABLE VII

SHOWING THE BREEDING BEHAVIOR OF TWO BLUE LEGHORN FEMALES, WHEN MATED WITH VARIOUS MALES OF KNOWN FACTORIAL COMPOSITION

♂	♀		Blue Splashed	Blue	Black	White
White Leghorn 117M $IIPP(rE)(rE)$	× 767 $iiPP(Re)(rE)$	Obtained Theoretical[5]	0 0	0 0	0 0	5 5
White Leghorn 117M $IIPP(rE)(rE)$	× 768 $iiPP(Re)(rE)$	Obtained Theoretical	0 0	0 0	0 0	11 11
Blue Andalusian 78M $PP(Re)(rE)$	× 767 $PP(Re)(rE)$	Obtained Theoretical	2 2	4 4	2 2	0 0
Blue Andalusian 78M $PP(Re)(rE)$	× 768 $PP(Re)(rE)$	Obtained Theoretical	3 3.5	7 7	4 3.5	0 0
White Plymouth Rock 155M $pp(rE)(rE)$	× 767 $PP(Re)(rE)$	Obtained Theoretical	0 0	20 22	24 22	0 0
White Wyandotte 192M $I^Pi^Ppp(rE)(rE)$	× 768 $i^Pi^PPP(Re)(rE)$	Obtained Theoretical	0 0	8 7	8 7	12 14
Blue {white Rock ♂ 155M / blue Leghorn ♀ 767} 255M × 767 $Pp(Re)(rE)$	× 767 $PP(Re)(rE)$	Obtained Theoretical	2 2.25	6 4.50	1 2.25	0 0
Black Andalusian 288M $PP(rE)(rE)$	× 768 $PP(Re)(rE)$	Obtained Theoretical	0 0	25 23.5	22 23.5	0 0

but two facts seem fairly evident. First, that 767 and 768 are alike with respect to the factors under discussion in this paper, and second, that they give no indication of being different in their make-up with respect to the factors R, E and P from pure-bred blue Andalusians.

The appearance of the blue offspring of 768 (which it will be recalled was very light) when mated with black or blue Andalusians, was such as to suggest the possibility that accessory factors, necessary for the production of blue of normal shade, were supplied by the Anda-

[5] The theoretical expectancies calculated as for blue Andalusians.

lusian males, though no attempt was made to isolate and identify them.

Since these blue Leghorns arose in an unpedigreed flock, their origin is conjectural. A plausible explanation seems to be that two individuals heterozygous for I, the dominant Leghorn factor described by Hadley (1913), which inhibits the production of pigment shown (also by Hadley, 1914) to be normally present in the white Leghorn, happened to mate and that at least one of them carried the factor R as a cryptomere. That white Leghorns may sometimes carry the factor seems to be shown by the fact that Dryden (1916, p. 67) secured blue chicks in an F_2 generation from a barred Plymouth Rock $\times$ white Leghorn cross. And further, in the course of the breeding operations reported in this paper, blues appeared in the progeny of a black Andalusian and a white crossbred, the latter being the product of a black Andalusian $\times$ white Leghorn cross. In both cases it appears that the factor R must have been brought in by the white Leghorn. This suggestion also involves the assumption that the white Leghorn carries the factor E. That this is the case is shown by the fact that in the F_2 from a blue-splashed Andalusian $\times$ white Leghorn cross, the details of which are reserved for later publication, both blacks and blues appeared.

XIII. The Problem of True-breeding Blues

The fact that the blue varieties of both Andalusians and Orpingtons as now constituted do not breed true is a matter of considerable importance to their breeders. It is a heavy handicap to both varieties. While one hundred per cent. of blues may in each case be secured by mating blue-splashed individuals with black, as a matter of practical breeding this mating is seldom made. This is owing to the fact that there are several more or less variable qualities of color for which rigid selection is practised which are not apparent in either the blue-splashed or blacks. The breeder therefore prefers to use for breeding purposes only those individuals which show the desired phenotypic condition, even though so doing necessitates the discarding of approximately half the offspring. While this leaves a comparatively small number of individuals, as compared with other breeds, upon which to practise selection, the blue Andalusian at least is bred in considerable numbers, thereby indicating its economic desirability and its attractiveness.

As was pointed out in the earlier paper (1918, p. 111) if R and E are not at identical loci on homologous chromosomes and crossover individuals were found which produce RE gametes, the problem of the constant-breeding blue would be solved.

The situation regarding black in rats may not be without its bearing in the present case. Black rats which bred true have been known for some time. Castle (1919) has, however, reported certain races of blacks which failed to breed true. This type of black was tested through several generations by Castle (1919), Ibsen (1920) and Dunn (1920). Blacks mated to blacks quite persistently produced whites, blacks and red-eyed yellows in the ratio of 1 to 2 to 1. Castle (1919) found one possible cross-over individual which died without being tested. Ibsen (1920) has so far failed to find any, and Dunn (1920) reports between one and two per cent. of cross-overs. These cross-overs, which were longer sought for and among larger numbers than has yet been possible with Andalusians, would appear to make it possible to synthesize a true breeding (*i.e.*, homozygous) black, from the line which has not been breeding true through a considerable number of generations.

It is also worth noting in this connection the possible bearing of Sturtevant's (1919) finding families of *Drosophila* carrying at least two definite factors in the second chromosome which almost completely inhibit crossing-over in the region contiguous to their loci.

If, however, after a long-continued search, it becomes increasingly evident that R and E are indeed allelomorphs, as originally suggested by Bateson and Punnett (1905), it was suggested (p. 113) that hope might be seen in the progressive selection of the darker, that is, more fully pigmented, blue-splashed individuals, there being considerable variation among the latter in this regard.

There is a further possibility which should not be overlooked, namely, that other factors might be found, perhaps in other breeds, which would act on black pigment to give the blue appearance on the one hand, or extend it to give self-colored individuals on the other. If duplicate factors for E or R should be found, a means of producing the long sought true-breeding blue would seem to be at hand. The fact that three factors are known which produce white in fowls lends emphasis to the possibility. It would seemingly make little difference in the ultimate outcome whether the new factor was linked to R and E, or was located on a different chromosome pair. In either case it would be possible to get a "self-coloring" and a "bluing" factor in the same gamete which, it appears, has so far not been done.

XIV. Summary

1. It has been shown that the development of black pigment in the blue-splashed, blue and black races of the Andalusian and Orpington breeds, and of black Langshans, depends upon the action of a dominant hereditary factor P, for which they are normally homozygous.

2. The allelomorph of P is p. Individuals homozygous for p are white, as in the white Wyandotte and white Plymouth Rock breeds.

3. The extension of black pigment to all feathers of the body, resulting, if no pattern factors are present, in self-colored individuals, depends upon a dominant factor E. This factor has been found in the Andalusian, Orpington, white Plymouth Rock, white Wyandotte and black Lang-

shan breeds. Some evidence is presented which indicates its presence in white Leghorns.

4. The blue appearance of blue and blue-splashed Andalusians and Orpingtons, is due to the arrangement and restriction of black pigment, the result of a dominant factor R. This factor has also been found in individuals of the white Wyandotte and white Leghorn breeds, though its presence is probably not usual in these breeds.

5. No individuals of the Andalusian, Orpington, white Plymouth Rock, white Wyandotte or black Langshan breeds have been found which did not carry R, E or both.

6. The mutual relations of R and E are such that they have never been found together in the same gamete. This indicates that they are allelomorphic, *i.e.*, occupy identical loci on homologous chromosomes, or, each is so closely linked to the recessive allelomorph of the other, (Re) and (rE), that crossing-over rarely, if ever, occurs.

7. No evidence of crossing-over between R and E has been found and the tentative conclusion must be in accord with that previously held, that R and E are allelomorphs.

8. Both R and E are independent of P in their hereditary behavior, though dependent upon its presence for their manifestation.

9. The cooperative influence of the ovary is necessary for a full expression of R in the regions of the neck, back and saddle.

10. On the basis of the evidence presented in the body of this paper the genetic formulæ of the breeds and varieties employed, with respect to the factors under observation, are usually as follows: blue-splashed Andalusians and Orpingtons $PP(Re)(Re)$; blue Andalusians and Orpingtons $PP(Re)(rE)$; black Andalusians, Orpingtons and Langshans $PP(rE)(rE)$; and white Plymouth Rocks and Wyandottes $pp(rE)(rE)$.

11. The possibility of the occurrence of factors which duplicate the somatic effects of R and E is pointed out, and the relation of this possibility to the production of constant-breeding blues briefly discussed.

XV. Bibliography

American Poultry Association.
>1919. The Plymouth Rock Standard and Breed Book. 432 pp. Published by The American Poultry Association.

Bateson, W., and R. C. Punnett.
>1905. Reports to the Evolution Committee of the Royal Society, II, pp. 99–119.

>1906. Reports to the Evolution Committee of the Royal Society, III, pp. 11–30.

Brown, Edward.
>1906. Races of Domestic Poultry. 234 pp. Published by Edward Arnold, London.

Castle, W. E.
>1919. Studies of Heredity in Rabbits, Rats and Mice. Carnegie Institution of Washington, Publication No. 288, pp. 56. 3 plates.

Cole, L. J., and F. J. Kelley.
>1919. Studies on Inheritance in Pigeons. III. Description and Linkage Relations of Two Sex-linked Characters. *Genetics*, Vol. 4, pp. 183–203.

Cole, L. J., and W. A. Lippincott.
>1919. The Relation of Plumage to Ovarian Condition in a Barred Plymouth Rock Pullet. *Biological Bulletin*, Vol. 36, pp. 167–183, 2 plates.

Davenport, C. B.
>1914. The Origin of Domestic Fowl. *Jour. of Heredity*, Vol. 5, pp. 313–315, 2 plates.

Dryden, Jas.
>1916. Poultry Breeding and Management. 402 pp. Published by Orange Judd Co., N. Y.

Dunn, L. C.
>1920. Linkage in Mice and Rats. *Genetics*, Vol. 5, pp. 325–343.

Goodale, H. D.
>1916. Gonadectomy in Relation to the Secondary Sexual Characters of Some Domestic Birds. Publication 243, Carnegie Institution of Washington, 52 pp., 7 plates.

>1918. Feminized Male Birds. *Genetics*, Vol. 3, pp. 276–295, 2 plates.

Hadley, P. B.
>1913. Studies on the Inheritance of Poultry I. The Constitution of the White Leghorn Breed. Rhode Island Agr. Exper. Sta. Bull. 155, pp. 149–216, 3 plates.

>1914. II. The Factor for Black Pigmentation in the White Leghorn Breed. Rhode Island Agr. Exper. Sta. Bull. 161, pp. 449–459. 1 plate.

Ibsen, H. L.
>1920. Linkage in Rats. Amer. Nat., Vol. 54, pp. 61–67.

Lippincott, W. A.
>1918a. The Case of the Blue Andalusian. Amer. Nat., Vol. 52, pp. 95–115.

1918b. Pedigreeing Poultry. Kansas Agr. Exper. Sta. Cir. 67, 16 pp.
1919. The Breed in Poultry and Pure Breeding. *Jour. of Heredity*, Vol. 10, pp. 71-79.

McGrew, T. F.
1901. American Breeds of Fowls. II. The Wyandotte. Bureau of Animal Industry Bull. 31, 30 pp.

Morgan, T. H., and H. D. Goodale.
1912. Sex-linked Inheritance in Poultry. *Annals of the N. Y. Academy of Sciences*, Vol. 22, pp. 113-133, plates 17-19.

Pearl, R.
1917. The Probable Error of a Mendelian Class Frequency. AMER. NAT., Vol. 51, pp. 144-156.

Platt, F. L.
1916. "Western Notes and Comment." *Reliable Poultry Journal*, Vol. 23, p. 65.

Punnett, R. C.
1919. Sex-linked Inheritance and Its Practical Application in the Breeding of Poultry. *Journal of the Board of Agriculture* (England and Wales), February, 1919. Summarized in Official Report of the International Poultry Conference, London, March 11 to 15, 1919, pp. 64, 65.

Rice, J. E., Nixon, C., and C. A. Rogers.
1908. The Molting of Fowls. Cornell Univ. Agr. Exper. Sta. Bull. 258, pp. 17-68.

Sturtevant, A. H.
1912. An Experiment Dealing with Sex Linkage in Fowls. *Jour. Exper. Zool.*, Vol. 12, pp. 409-518.
1919. Inherited Linkage Variations in the Second Chromosome. Carnegie Institution of Washington, Pub. 278. Part III, pp. 305-341.

www.ingramcontent.com/pod-product-compliance
Lightning Source LLC
Chambersburg PA
CBHW062336220526
45469CB00008B/2733